Man and Forest Series
No. 5

General Editors

Klaus Seeland
Franz Schmithüsen

A joint research venture of the

Council of Cultural Growth and Cultural Relations, Cuttack

and

Chair of Forest Policy and Forest Economics, Swiss Federal Institute of Technology (ETH), Zurich

in co-operation with

Deutsche Gesellschaft für Technische Zusammenarbeit (GTZ)
Tropenökologisches Begleitprogramm (TOEB)
(Tropical Ecology Support Programme)

Man and Forest Series

General Editors: Klaus Seeland; Franz Schmithüsen

The Man and Forest Series consists of monographs and readers on the use and management of forests and an account of forest lifestyles in various regions of South Asia. It aims to provide the findings of an eight year research project (1992-99) to the public. The volumes in this series are the outcome of a cross-cultural research network between the Chair of Forest Policy and Forest Economics at the Swiss Federal Institute of Technology (ETH), the Deutsche Gesellschaft für Technische Zusammenarbeit (GTZ), and Non-Government as well as Government Organisations in India, Nepal and Bhutan. The series welcomes contributions covering relevant issues in the man-forest interface.

Indigenous Knowledge, Forest Management and Forest Policy in South Asia

Indigenous Knowledge, Forest Management and Forest Policy in South Asia

(Proceedings of an International Seminar held in Kathmandu, Nepal in 1998)

Edited by

Klaus Seeland
Franz Schmithüsen

D.K. Printworld (P) Ltd.
New Delhi

Cataloging in Publication Data — DK

Indigenous Knowledge, forest management and forest policy in South Asia: proceedings of an international seminar held in Kathmandu, Nepal in 1998 / edited by Klaus Seeland, Franz Schmithüsen.
(Man and forest series; no. 5).
Includes bibliographical references.
Includes index.

1. Indigenous peoples — South Asia — Congresses. 2. Forest management — South Asia — Citizen participation — Congresses. 3. Forest policy — South Asia — Congresses. 4. Tribes — India — Congresses. I. Seeland, Klaus, 1952 – II. Schmithüsen, Franz, 1940 – III. Series: Man and forest series; no. 5.

ISBN 81-246-0222-0

First published in India in 2003

Supported by: Tropical Ecology Support Programme (TOEB) Deutsche Gesellschaft für Technische Zusammenarbeit (GTZ)

Published and printed by:
D.K. Printworld (P) Ltd.
Regd. office : '*Sri Kunj*', F-52, Bali Nagar
New Delhi - 110 015
Phones : (011) 2545-3975; 2546-6019
Fax : (011) 2546-5926
E-mail: dkprintworld@vsnl.net
Web: www.dkprint.com

Preface

In the research network 'Man & Forest', to which most of the contributions to this volume refer, many institutions, individuals, scientific researchers and administrators have been involved. Other contributions to this book have included inspiring additions to what the organisers of the Kathmandu workshop in 1998 focused on and what the title of this book stands for: *Indigenous Knowledge, Forest Management and Forest Policy in South Asia*. Many of the contributors shared a keen interest in the meaning and relevance of these topics, especially in the Nepalese context. Some were involved as practitioners in forest-oriented development co-operation. Others were members of local non-governmental organisations, involved in community forestry or village development programmes.

The proceedings of the international seminar in this volume reflect the state-of-the-art research activities that evolved from the first conference in this research network which was held in March 1995 in New Delhi. The contributions to this earlier international seminar were published as volume 1 in the 'Man and Forest Series' under the title 'Man in the Forest – Local Knowledge and Sustainable Management of Forests and Natural Resources in Tribal Communities in India'. In the second phase, the research venture Man & Forest (1997-99) sought to communicate indigenous knowledge of forest and natural resource use management in a wider Himalayan perspective to a growing scientific community which takes an interest in local and indigenous knowledge. More partners joined the network who either perceived indigenous knowledge as a development tool or wanted to contribute to the preservation of tribal cultural identity and know-how in an age of standardised modern approaches and rapid social change.

This book aims to meet the challenges of the present by avoiding a glorified or romanticised image of the past which has never existed. Local indigenous knowledge did not solve all problems of traditional societies in the past. The reasons for coming back to indigenous knowledge and self-reliant forest and natural resource management is at least twofold among many other reasons. It is essential in times of administrational deregulation to have collective initiatives at the grassroots level with a high degree of self-organisation, self-regulation and locally-accepted patterns for securing the livelihood of independent village communities. Independent in the sense that they are not eager to become clients of governmental social services, but continue to bear responsibility for themselves and for their immediate environment.

Although most of the contributions to this book are on India this includes its wider northern borderlands. Tribal lifestyles and agrarian communities in remote forest areas were, particularly after Indian Independence, industrialisation and the emerging nation state, perceived as an obstacle to modernisation. Community development programmes and the promotion of Panchayati Raj (decentralised, local village government) were approaches designed to adapt the diversity of Indian cultures to western notions of economical advancement. Community development programmes defined traditional social structures as a development problem. Traditional lifestyles, wherever they mattered practically in the economic development process, seemed to hamper the building of a modern India. Cultural diversity was almost equated with folklore reminiscent of the unification policy that was meant to encompass this vast patchwork of nearly independent village civilisations during the 1950s and 1960s. Gandhians, and later on the Indian green movement that emerged in the 1980s, had, or claimed to have, a more Indian vision of what authentic development should be all about. After the Green Revolution had failed largely due to its negative impact on the environment and social equity in rural societies, it became obvious that the prize of modernisation had to be paid by the environment and the rural poor. Indian cultural values, going back to the sources of the so-called *Little Tradition* of the tribal and local religious cults, and to the *Great Tradition* of Hinduism, Islam, Buddhism and Jainism, became increasingly appealing to give development and social change a more humane perspective. The survival of those who live a marginal life and exist on what Nature provides them with was put on the political agenda.

Indigenous knowledge of forests and the management of local natural resources have a tremendous impact on the social development of rural areas, not only in India but in the adjacent countries of Nepal and Bhutan as well. As reflected in the contributions to this volume, South Asian rural communities share much of a common legacy in particular as far as their exposure to (primitive) living conditions close to Nature is concerned. A common feature of most communities referred to in these proceedings is their social cohesion within their respective village community groups and, more often than not, readiness to participate in the informal or formally institutionalised management of forests and forest products.

One of the most important aspects of forest life in the wider Himalayan context is that human needs are almost entirely interwoven with Nature, be it physically, socially or spiritually. Once they have become disconnected from the direct satisfaction of their needs with natural products taken from the forest and traded as market commodities, there is often an alienation that may disrupt forest dwelling communities from their attachment with and responsibility for their surroundings. The removal of rights, entitlements and concessions to use forest products due to the nationalisation of forests in India (1865), Nepal (1957) and Bhutan (1969) changed the lifestyle of the major part of the rural population in these countries. To manage forests in a self-reliant way, in accord with the national demand for forest products, and in view of the forests the role for the preservation of soil fertility and forests play climatic conditions, is the goal of social and community forestry in the wake of joint forest management activities. These activities are designed to bring local users, non-governmental organisations and officials of the national forest services to terms with one another. The success or failure of these forest policies depends largely on whether the respective country's policy is accountable to the users who are often sceptical about the credibility of government induced programmes. Yet users are never a homogeneous group. On the contrary, they are highly heterogeneous with respect to their demands, ambitions, expectations and interests.

Future forest policies in developing nations (and this does not only take into account the countries which are referred to in this volume) will in all likelihood be characterised by an appreciation the cultural foundations of forest use and the management of renewable natural resources by the state administration. New deals will have to be negotiated between the local and the national level. In this process, global goals of nature conservation may interfere with national political autonomy and favour local initiatives. International co-operation between foreign donor agencies and national or local NGOs has demonstrated this over the last two decades. Forest policy will thus increasingly become an area of globally induced social policy with regard to the sustainable management of natural resources based on a diversity of knowledge cultures.

The editors gratefully acknowledge the help of Terence Hay-Edie in the editing process. The photographs for this volume have been provided by H.N. Das, Chr. Giesch, M.K. Jena, Chr. Rosset and K. Seeland.

Klaus Seeland
Franz Schmithüsen

Contents

Preface vii

List of Abbreviations xiii

Contributors xvii

1. Whispers of the Wild 1
   — Girija Bhusan Patnaik

2. Indigenous Knowledge and Tribal Forest Life in Orissa 13
   — Klaus Seeland

3. Man-Nature Relationship of the Juang of Juangpirh (Kendujhar District, Orissa) 27
   — Nityananda Patnaik

4. Biodiversity in Juangpirh, Orissa 55
   — B.P. Choudhury

5. Food, Folk and Forest 67
   — Alok Rath

6. Economical and Biological Potential of Non-Timber Forest Products (NTFP) in the Similipal Hills of Orissa 81
   — Mihir Kumar Jena

7. Forest Use among some Tribes of Orissa 105
   — Kamala Kumari Patnaik

8. Joint Forest Management — Merger of two Traditions? 121
   — Nadja Ottiger

9. What alternative would Conserve Forests Better? Empirical Studies on Joint Forest Management in Orissa 145
   — Padmini Pathi

10. People's Institutions and Joint Forest Management – Possibilities and Problems in the Context of Rajasthan (India) 173
— S.N. Bhise
— A.S. Mehta

11. Agroforestry Development and Management through Human Resource Development at the Grassroot Level 195
— Laj Pal Bharara

12. Community Forestry in the Terai: The Churia Forest Development Project in Nepal 211
— Jürgen Fichtenau

13. Participatory Management of Natural Resources with Forest User Groups — A socio-cultural study in the mid-hills of Nepal 235
— Christian Rosset
— Klaus Seeland

14. Environmental Perception and Forest Management among the Tamang of Southern Kavre 265
— Buddhi Tamang
— Amrit Yonjan

15. Traditional use of Blue Pine (*Pinus Wallichiana*) and its impact on the Forest Structure in Central Bhutan 289
— Christina Giesch

16. Simple Participatory Forest Inventory: A Key Factor for Sustainable Forest Management 311
— Robin aus der Beek

Visuals 327

Glossary 367

Index 369

List of Abbreviations

CF	:	Community forestry
CFM	:	Community forest management
CFUG	:	Community forest user group
ChFDP	:	Churia Forest Development Project
CHIPKO	:	Hugging-the-trees movement in Uttarakhand (north India)
CPR	:	Common property regime
dbh	:	Diameter at breast height
DDC	:	District Development Council
DFO	:	Divisional Forest Officer (in India)/District Forest Office(r) (in Nepal)
DMP	:	Decision making process
EC	:	Executive committee
FD	:	Forest Department
FECOFUN	:	Federation of Community Forest User Groups in Nepal
FPC	:	Forest Protection Committee
FUG	:	Forest User Group
GDP	:	Gross Domestic Product
GIS	:	Geographical Information System
GPS	:	Geographical Positioning System

GTZ	:	Gesellschaft für Technische Zusammenarbeit
HMG/HMGN	:	His Majesty's Government of Nepal
IK	:	Indigenous Knowledge
IRs	:	Indian Rupees
ITDA	:	Integrated Tribal Development Agency
JFM	:	Joint Forest Management
LAMPS	:	Large and Multipurpose Societies
LPP	:	less privileged people
MFP	:	Minor Forest Produce
MoEF	:	Ministry of Environment and Forests
NCA	:	National Commission on Agriculture
NGO	:	Non-Governmental Organisation
NRs	:	Nepalese Rupees
NSCFP	:	Nepal-Swiss Community Forestry Project
NTFP	:	Non-Timber Forest Produce
NWDB	:	National Wasteland Development Board
OP	:	Operational Plan
ORMAS	:	Orissa Rural Development and Marketing Society
PCF	:	Programme of Community Forestry
PRA/RRA	:	Participatory Rural Appraisal/Rapid Rural Appraisal
PSU	:	Project Support Unit
PWG	:	Armed extremist group
RGOB	:	Royal Government of Bhutan
S.C.	:	Scheduled Caste
SLC	:	School Leaving Certificate

S.T.	:	Scheduled Tribe
VDC	:	Village Development Committee
VFC	:	Village Forest Committee
VFPC	:	Village Forest Protection Committee
VLW	:	Village Level Worker
VSS	:	Vana Samrakshana Samiti (Forest Protection Committee)

Contributors

aus der Beek, Robin, Forestry advisor, Nepal-Swiss Community Forestry Project, Kathmandu, Nepal

Bharara, Laj Pal, Principal Scientist (retd.), Centre of Arid Zone Research Institute, Jodhpur, Rajasthan

Bhise, S.N., Forestry advisor (retd.), Seva Mandir, Udaipur, Rajasthan

Choudhury, B.P., Professor of Botany and advisor, Social Science and Development Research Institute, Bhubaneswar, Orissa

Fichtenau, Jürgen, Forestry advisor, Nepal-German Churia Forest Development Project, Kathmandu, Nepal

Giesch, Christina, Technical advisor, Social Forestry Support Programme, Hanoi, Vietnam

Jena, Mihir Kumar, Director, AEI Research and Consultancy (P) Ltd., Bhubaneswar, Orissa

Mehta, Ajay S., Executive Director, National Foundation for India, New Delhi

Ottiger, Nadja, Assistant, Ethnological Institute, University of Zurich, Switzerland

Pathi, Padmini, Ph.D. scholar, Utkal University, Bhubaneswar

Patnaik, Girija Bhusan, General Secretary, Council of Cultural Growth and Cultural Relations, The Universe, Cuttack (Orissa)

Patnaik, Kamala Kumari, Professor of Botany (retd.) and member of the Orissa Public Service Commission, Cuttack (Orissa)

Patnaik, Nityananda, Director, Social Science and Development Research Institute, Bhubaneswar, Orissa

Rath, Alok, Senior Researcher, Social Science and Development Research Institute, Bhubaneswar, Orissa

Rosset, Christian, Assistant, Chair of Silviculture, Swiss Federal Institute of Technology (ETHZ) Zurich, Switzerland

Seeland, Klaus, Senior Lecturer, Chair of Forest Policy and Forest Economics, Swiss Federal Institute of Technology (ETHZ) Zurich, Switzerland and Reader in Sociology at the University of Konstanz, Federal Republic of Germany

Tamang, Buddhi, Member of SAGUN, a Nepalese NGO based in Kathmandu, Nepal

Yonjan, Amrit, Member of SAGUN, a Nepalese NGO based in Kathmandu, Nepal

1

Whispers of the Wild

Girija Bhusan Patnaik

THE whistling wind whispers. The deers run intoxicated with the gentle breeze, passing through the trees in the Forest. The cave echoes the roars of the tigress searching for her mate. The wild flowers blossom infusing the air with their enchanting sweet smell. The smoke rises from the Kondh or Saora kitchens whirling into the sky, like the hairs of a damsel floating in the air. And they dance and sing together to greet the new season, be it *karba* or *karma.* They are born in the forests, grown in the forests and also die in the forests. Forests govern their culture, their economy, their world-view and their performing arts. After death, their *duma* (souls) hovers round and plays a role to mould the lives of the succeeding generation. They talk to the spirits through the *disaris* (the middle man, the witch doctor) and pray for good rains, good crops, cure of diseases and well-being for all.

Life still remains within mountain ranges, rising high up into the sky. Their *world* moves within the mountains and beyond it. . . . The Dongaria Kondh looks askance, his eyes trying to grasp the world outside. He is not able to conceive a world outside, though intruders have entered his 'land' and exploited him. According to his world-view he originated from below these mountains alone. The exploiters in the earlier days were the *mahajan*s, the money-lenders, mainly belonging to the castes of brahmanas and sundhis, who became synonymous with the exploiters. And today ? The scenario has changed, serpentine tarred roads bring in new types of exploiters and exploits, their living conditions like the sky above remain the same.

The anti-colonial, freedom fighters, as well as the makers of the Indian Constitution, appreciated that without the development of the tribals the fruits of freedom would not be realised in full. Tribals still constitute a large population. They constitute 22 per cent of the population in Orissa, for example, and constitute a sizeable proportion of the population in other States like Tripura, Bihar, Madhya Pradesh and form a majority in the North- East Hill States of India.

Their role in the Freedom Movement is significant. From the early days, they raised not only the banner of revolt against British imperialism, but also against local tyrants and rulers who collaborated in the rule of exploitation.

The historic Santal revolts in Singhbhum led by the legendary Birsa Munda; the Bhuyans and Juang of Kendujhar led by Dharanidhar Bhuyan; the Kondh of Angul led by Anu Kondh and the rebellion in Dasapalla; are the torch-bearers of the Freedom Movement. The rebellion of the tribes in the Koraput District, led by Laxman Nayak, the tribal who went to the gallows in the 1942 Quit India Movement, is a beacon for the future.

The illiterate tribals knew more than their educated non-tribals that the fight against tyranny, be it in-land or foreign, cannot be victorious without being linked with the non-tribals. The struggle had to be unified, which is now forgotten. The heroic struggles of Birsa Munda is a clear illustration of the unified struggle they envisaged.

Hence, the founders of the Constitution of India while enshrining the Fundamental Rights also envisaged the 'Directive Principles of State Policy' for the development of tribals and scheduled castes with provisions to create reservations for a period of ten years. They thought that this period of ten years would be sufficient, but alas ! even after 50 years of Indian Independence, the tribals remain almost where they were, with no qualitative changes in spite of government money for their development. The policy of reservation continues, as if in perpetuity.

Government sources admit that nearly 65 per cent of the money for the tribals, is pilfered by middlemen: a coterie consisting of a small of tribals and the rest non-tribals. The non-official estimates put it higher even higher. The result is that a group among the tribals have come to be named the 'New Brahmanas' (synonymous with exploiters) who monopolise everything for themselves and their families, including the elective posts.

In the meantime the traditional forests or the natural habitats are being denuded. It is estimated that India needs annually 25 to 35 crore cubic meter of timber for firewood, while the production is limited to 4.5 to 5 crore of cubic meter. How to compensate this? Alternative means of fuel is confined to a small section of people. Similarly, for a country forest protection needs a coverage of at least of the land, one-third. In India, it has reached alarmingly low to 21 per cent, while in Orissa which had the proud privilege of impregnable forests the percentage has gone down to 14 per cent only. The remote-sensing on which these calculations are made, often records shrubs and bushes as forests and hence the 14 per cent coverage is also doubtful. In India some 3,000 species of plant, out of about 15,000 are extinct. Similarly 70 out of 322 mammal species, 42 out of 1,175 species of bird, 322 out of 399 reptile species are either extinct or on the verge of extinction. This has created a serious threat to the regional ecosystem.

Ask the so-called civilised man, he will put the blame on the shifting cultivation of the tribals. But ask the tribals for the sake of answering. They will point to the patches of forests which they have used for shifting cultivation with stumps of trees kept intact to regrow in future. But what happened to other forests ? Hills after hills, once covered with dense forests, have been blasted out for Iron and Manganese ores or Bauxite without the minimum care for re-plantation of the area. Predators of forests and forest animals have plundered the forests in connivance with a group of corrupt forest officials, police and with the patronisation of politicians. The honest ones have been punished or have suffered in ignominy. One is not surprised if some of these political patrons have become member of parliament, legislative assemblies or even ministers.

Dews still drop from the trees as tears of the tribals whose livelihood is lost. The intoxicating essence of the *mahua* flower still thickens the air, the barking of the deer greeing into the distance, the melodies and the rhythms of the tribals still echo from the denuded, barren hills and mountains, and the tribals are finally searching for alternative means of livelihood.

A division of opinion regarding the development of tribals creates debate. Some people want the tribals to remain at their present state of illiteracy, poverty and nudity which will enable them to *study* their primordial lifestyle and to take their nude photographs. But their research is library-oriented, and cannot be verified, simply because who cares and who dares to verify them?

There are others who think that tribals should be brought into the mainstream of civilisation and there is no need for their special identification. Missionaries fall into this category. But there is a third trend which envisages that the tribal's development should be handled with care by the people who love their culture. There should not be a feeling of a 'benefactor' and 'beneficiaries', which creates a gap between the two. The tribals have their own culture, tradition, language and world-view and these should not be polluted, diluted or disturbed. The tribals belong to the forests and the forests also belong to them. Their lifestyle and living should be developed in a manner so that their education, economic development should to built around their environment, in particular the forests.

The tribals, especially the women, are still far behind in formal education, which is a must for their development. It should be conducted so that they learn from their own surroundings and environment. Tribal women are often equal with their menfolk and sometimes more as in the case of the Saoras tribe. Hence, education of women is important, yet nothing tangible or substantial has been done in this respect. The second important conclusion we have found is that alternative means of livelihood should be taught alongside vocational education so that tribals should not depend only on shifting cultivation or on the forest for their food and firewood. But what type of vocational education they should get?

They live inside the forests, in hamlets known as 'Forest Villages'. People living there earlier were given rights over the minor forest produce including firewood. In turn they used to protect the forests, even against forest fire under the supervision of the Forest Department. It was a Joint Forest Management (JFM) in its nascent form. This was abandoned/scrapped and the management came directly under the Forest Department (FD). The people dwelling in the forest or in the vicinity remained aloof and their emotional attachment to the forests was lost in the course of time. In the National Forest Policy 1988, a clear stipulation was made for the involvement of people in the activity of forest conservation. The Government of Orissa took another step in that direction by formulating the Orissa Village Forest Rules, 1985, and then issuing a resolution in 1988 whereby the task of the protection and management of forests was assigned to the fringe villages. Subsequently, large numbers of forest protection committees were formed as envisaged in the resolution. However, for various reasons, this task lost its momentum.

The resolution on 'Joint Forest Management, 1993' gave the desired impetus to the task of Forest Protection and Management through popular participation by specifying the sharing of usufructuary benefits. In 1996, the Government of Orissa passed another resolution on JFM, supplementary to the resolution passed in 1993, wherein it has been envisaged to provide proprietary right to the people of the adjoining villages over the forests which are being protected by them. The draft guideline for the implementation of the 1996 resolution has been formulated by a working group constituted specifically for the purpose, and a Government decision is awaited.

With recent summer temperatures rising, unprecedented levels and sun-stroke taking a heavy toll of more than 2000 lives in Orissa (as per the official records), environmental awareness has grown, both in the cities and villages leading to a spree to plant more and more trees. It is now up to the Government it as to how will rise to the occasion to lead the people.

The type of vocational education to be given to the tribals is yet to be decided. Forest products now have national and international markets. Vocational education should impart training in the scientific processing of these products so that they become exportable. Tribals today are almost forced to grow fruits, etc., which they neither consume, nor know how to market. They should be encouraged to grow products which they consume and efforts should be made to export the other ones to get a suitable earning. As far as possible, middlemen system should not be allowed to make benefits out of this. Tribals now produce plantains, pineapples, oranges, the benefit of which goes to the middlemen, mainly non-tribals, who purchase them in a nominal price taking advantage of a system of advancing money at the time of cultivation. The result is that the tribals remain poor to the benefit of the middlemen who have clout with the politicians in power and the food processing industries in the tribal districts remain 'locked and closed'. Take the example of hill broom called *'phul jhadu'* which millions purchase at a throw-away price from the tribals and export to foreign countries with small additions like a plastic cover at a price more than 30 times higher than the purchasing price. The tribals of Niyamgiri Hill remain as poor as they were.

The tribals should be brought together in co-operatives with a real co-operative spirit and the Government who should see that their products are also marketed have effectively. The dwelling place of tribals is incidentally rich in minerals. Large-scale industries are coming up or have come up in these areas, and tribals have lost their land and the forests. Without effective utilisation, the compensations paid to them are wasted. Their land was taken for industrial and commercial purposes, but in many cases the compensation is paid for agricultural land. This has been resented and in some cases resisted by the tribals.

The tribals, who lost their hearth and home, can be trained to enable them to earn their own livelihood by working as masons, electric fitters, as pump operators, etc. They earn their livelihood in a new way and the stress on forests is reduced. They earn their livelihood by alternative means, so that they do not resort to shifting cultivation.

There are three important aspects to our work:

1. We have organised a residential school for tribal girls in the village of Banjikusuma in Rayagada District, where 50 live-in students are getting formal education from Class-I to Class-V. We plan to have five more residential schools in the same District, if we can get financial assistance from the Government of India or any other agencies.
2. We have started a pilot project in the village of Dangsorada in Rayagada District where tribals, especially women, are brought under a co-operative society and are engaged in making leaf plates and cups. But this has run into difficulties because of the lack of governmental co-operation in marketing the products in the cities situated a long distance away. The Government of Orissa had formed an organisation called ORMAS for this purpose but this has almost been closed down because millions of rupees were

pilfered by non-tribal ministers and officers. Hardly anybody has been punished, though found guilty, and the organisation went bankrupt.

3. We plan to have three or more Vocational Education centres, one at Rayagada, one at Phulbani in Kondhamal District and one in the vicinity of Bhubaneswar where tribals, rural youth and slum dwellers co-exist.

We plan to impart training by which such groups learn the processing of forest products and the art of cooking their food with less firewood. We are trying to create a fusion of the tribals and poor of the non-tribals so that conflicts amongst them, a regular phenomenon, is eradicated or mitigated.

We need financial support to implement the project. We have applied for grants, both domestic and foreign, governmental and non-governmental. Some have been pending for years, probably owing to the notion that the tribals who have waited for centuries can wait for a few years. At other places, people sitting in air-conditioned rooms in Delhi have assumed for themselves the role of saviours and experts of the tribals and hence decided that the proposals or plans are not 'practical' or 'feasible' without caring to look at the small experiments already made. Others still, have given funds elsewhere.

They do not hear the tremors, the rumblings. They are not able to hear the sound of the blooming flowers, the smell of the fragrance electrifying the air, or feel the vibration of the flute played over the hills creating waves in the air. They feel safe in their cushioned chairs, and the coolness of the air-conditioner has breezed their mind. They have decided the fate of the tribals for centuries and they think they will continue to do it for ages to come.

But the time has started changing. The drums of the tribals are beating a new rhythm, the rhythm of struggles, the wayward tune of the flute still calls for the beloved not only for bliss but also to fight together. Hear, hear them and you will hear the beginning of the change.

We appeal to all of you, dear friends, to use your goodwill with organisations, institutes and foundations to come to our help. It is not the development of, or for, the tribals, it is our own struggle to live. A large proportion of mankind living below the poverty line in perpetuity is a threat to humanity. The destruction of tropical forests will destroy, not only a source of poetry or painting humanity itself will be threatened.

Hills of Ramgiri, the clouds hover over them like the apron of a young damsel. The flying ducks clasping their wings on the advent of the monsoon created not only the immortal poems of the *Meghaduta* of Kalidas, but also kept the ecological balance for the entire people. It is time to ponder if everything aesthetic should be butchered at the altar of utilitarianism or should there be a careful balance. Who will keep the balance if we forget the tribals?

And finally what happens to the valuable medicinal plants, which are going to be extinct, their identification lost and so also their uses? The twenty-first century has a great demand for alternative medicines. We are all aware of it, yet we are a long way from identifying the herbal plants, in recording their uses, and preserving them in their natural environment.

With no support, financial or otherwise, we have tried to play our modest role as far as possible. The Co-Director of the project 'Man and Forest' in Orissa is a botanist and a specialist on herbal plants and uses. She has already published two volumes of books in Oriya for the immediate use of the local people, and the third volume is in progress. They have not been printed and published in English because of lack of resources. But a concerted and well-planned action is yet to begin. We have suitable personnel but do not have financial support or strength. The tribals know plants for family-planning with no side effects. They use plants and their extracts against Bronchial Asthma, a dreadful disease in environmentally polluted cities. They know of plants which cure Arthritis. The list is endless. A regular study, with a team of researchers, for the identification of plants and their uses, their preservation in herbariums, and then replanting in their natural environment is an immediate necessity, followed by multi-departmental experiments. This there is an international demand for task cannot be carried out without support and we hope that since alternative medicines, the help will also be global.

Our research on 'Man and Forest' is not a barren research for research's sake. There certainly lies a larger, wider, deeper purpose for it. We are not only academics but also social activists. We have a mission in our eyes. Our targets extend for into the twenty-first century. We did not go for studies alone but also to find solutions.

Our team lived with the tribals, ate with them, sang with them, danced with them. Our scholars started living with them, learning their language, and getting acquainted with their culture. They provided us with some very interesting conclusions. They heard the trumpets of elephants near the hut in which they lived, their slumber was broken by the roar of the tiger, making the hairs stand on their roots. They walked down the serpentine, sleepy and windy roads, one after the other, in a queue, means tail identifying the method of movement of the Lanjia Saora. The steep hill rose on one side and the slippery slopes on the other with a passage for hardly one person to walk. Yet they marched with a mission :

> The woods (forests) are lovely,
>   dark and deep,
> But I (we) have promises to keep.
> And miles to go,
>   before I (we) sleep,
> Before I (we) sleep.

The beautiful, curly hairs of the young research scholar, is now unkempt, undressed for months. The dusty locks of her hair is loosely tied at the back of her head. Her calfs are aching, her toes have developed ulcers. The refrains of the flute dying out in the air enchant her calling her towards a larger mission. The tiger and the elephant on the one side, and Naxalites or the armed PWG group on the other, she marches without fear her head held high. To quote Rabindranath Tagore, 'to the heaven of freedom let god my country awake'. She marches along with her associates and comrades with a dream that some day, probably in the twenty-first century, her findings will be implemented not only for the tribals but also for the larger society. She is not afraid of the 'gentleman of the forest', as the tiger is called, nor of the armed Naxalites (among whom there are ladies too) because her larger interest combines with the others where the parallel lines meet. The social aims implemented will be a bigger and more coveted award than a *Padmabhusan* or *Padmashree* for her. Her dress has become dirty, hair unkempt, scratches have occurred in the dried up skins, dried as much in the summer blizzard as in the sera winter, winter that walks from the bushes to the hut like the tiger but her mind and heart becoming more and more clear, cleansed by the simplicity of the tribals and the fragrance of the virgin nature, their land after the first showers of monsoon, its trees blossoming at the dawn of spring and the cool water of its rivulets jumping from one stone to the other and creating a nostalgic music.

The tribals keep the scholars young in spirit though they have grown in age. Time has become infinite for them as for the tribals. Nothing remarkable has taken place, no Christ has appeared, no Gautam has attained *baudhitva*, no Vivekananda has marched through the forests, to record time.

Sometimes a bell in the church of the Christian tribals, the Ajans from a stray Mosque, and the blowing of a conch-shell in distant temples, break their dreams. The sound of the conch-shell echoes in them like the sounds of *panchajanya* by Srikrishna, in the war of *Mahabharata*, heralding a New Age or *yuga*. Time has stood still for them, years and months passing by smoothly, without any sound, like the day meeting its crimson end turning into night, without any sound. Time has expanded into the starry sky and beyond the mountains.

Orissa has 62 tribes officially recorded. One more tribe Paharia has been identified. The researchers dream to cover all the tribes or at least a bulk of them, opening a new vista, a new horizon.

Friends, we are on the threshold of the twenty-first century. The twentieth century, probably the most important century in the history of mankind, is coming to an end.

And we are standing at a crossroads. The world has become very small. Depletion of tropical forests in Brazil threatens Europe, leakage in the Ozone layer over Greenland, threatens Asia, the blast in the Atomic Reactor in Chernobyl in the erstwhile USSR poisons vegetables in Italy.

Let us recall a quotation of the Blind Oriya Tribal poet Bhima Bhoi, born in Rairakhol in the nineteenth century, known for its impregnable forests. The poet wrote the immortal lines:

> who sees the sorrows
>   of the people,
> traversing on the earth,
> O, God, Let my soul dwell in
>   the hell,

but let the world
be salvaged

This hymn still haunts the tribals and also haunts our scholars. The final message comes, the person is unimportant, what is important is the common good and welfare. Hence, we may either live or perish, we must choose between the two. Forests are a vital component. And the choice is ours, yours and mine.

2

Indigenous Knowledge and Tribal Forest Life in Orissa

Klaus Seeland

Indian forests — A vital aspect of Indian culture

EVERY venture into Indian culture will come across the forests of India somewhere. In the dawn of the Vedic age, forests are referred to as the cradle of civilisation and culture (Das 1988). Any attempt to assess the situation of India's forests today must similarly reflect the role of man and forest in a perspective of sustainable use and management of renewable natural resources for the twenty-first century. India's wealth and beauty as a predominantly agrarian country, its soil fertility and availability of water, are among many other aspects linked with its forest cover. The ecological survival of the Indian subcontinent is ultimately linked to the preservation of forests. In particular, a substantial part of the 80 million tribal people of India are still dwelling in the forests or make a good deal of living from them. India's environmental history (Gadgil/Guha 1993) shows that forest habitats, even under modern conditions of deteriorating forest cover, have a lasting continuity. Indian culture has developed under the specific conditions that characterise its environment, and environmental degradation is a term not unknown to Hindu cosmology as a concept of decay in the *kaliyuga*, the age of emptiness and deterioration.

The persistence of forest cultures has a remarkable influence on the foundation of the Indian green movement and should not be underestimated (Shiva 1991). The discourse on tribal rights and forest legislation (Fernandes 1996; Sharma 1995) has had a tremendous effect, not only on the Indian public, but has for a long time influenced the international debate concerning intellectual property rights and the political empowerment of forest-dwelling communities. Among many other tribal communities in India, a relevant section of the 63 tribes of Orissa represent this forest culture. Hitherto, only some of them have been selected for inquiry. With such an investigation, the rethinking of a still living aspect of Indian tradition and culture shows what role the forests of India are to play today and in the future.

Access to the forest world

Research on Indian forest culture first of all requires appropriate access to a life world so often referred to as primitive. How are the forests and other natural resources perceived by some of Orissa's tribal groups? What can be said about local or indigenous knowledge of forests and forest management? What can be learned from their rationale of forest use and management, and could they be integrated into modern development concepts of sustainable forest management? These questions were selected by a research project, which was supported financially by the governments of the Federal Republic of Germany and Switzerland. Some of the contributions to this volume are an outcome of research work that was done in, or in connection with, this venture. One of the goals of the research network including Indian voluntary agencies, a Nepalese non-governmental organisation, and research scholars working along similar lines, was to evaluate if and how these perceptions and management patterns could become a basis for a culturally and ecologically appropriate forest management in co-operation with the concerned forest dwellers. A prerequisite to obtain relevant insights into the livelihood patterns of Orissan tribal communities was to try to understand their living conditions and world-views.

How to understand the tribal world?

Those who take an interest in the life world of forest-dwelling tribes in Orissa should be aware of the paramount spirituality and religiousness that shapes their everyday life, their perception, and how they order environmental phenomena which represent and connect all spheres of their material and immaterial surroundings. Their world is defined by cosmogonic myths (see Elwin 1954, or Seeland *et al*. 1995, Appendix 1.1(A)) and even in a transitional phase between their traditional lifestyles and all-pervading modernity, they refer them to as a basic layer of their culture. In its most fundamental aspects, their lifestyles are an almost eternal repetition that runs along the lines of myths and ancestral cults, but are also adapted to the demands of either modern or non-tribal Hindu lifestyles wherever it appears to be either advantageous or unavoidable. In the tribals' minds things are not easily changed by the living. The power of ancestors and tradition is, besides interventions from outside (i.e. the state or civil society), the dominant ordering power that structures a forest dweller's life. There can be no understanding of forest lifestyles, if the social meaning of a tribe's surroundings are not seen and interpreted on the basis of this background.

Methodology

The research network which inquired into the Orissan forest cultures started with an approach which left it open as far as possible concerning which phenomena would be relevant for the research, such as what forest is, and what it means in the tribals' eyes. To approach the tribal forest world in the broadest possible perspective, the methodology of 'hermeneutic contexts' was chosen. Contexts are configurations of aspects which are related to each other and represent what ultimately makes up their cultural meaning. All possible contexts are methodologically imagined to represent the orbit of a local culture. These contexts refer to other contexts which are linked again to other configurations. These linkages are facilitated by language, such as by the use of metaphors and polysemic meanings for words or key symbols representing certain plants, animals or human relations with them. Contexts are comprehensive units in which culturally meaningful compositions of relevant material, or immaterial relations, are constituted. The authenticity of data, defined by the context-methodology, is represented in the respective contexts they constitute. These contexts refer to each other and ultimately form a cultural whole. A hermeneutic understanding of trees, forest plants, animals and humans as related phenomena is thus possible for a scientific description and a cross-cultural comparison between several tribal communities within that region.

The scope of implementation of indigenous knowledge

Contexts and configurations can only be investigated with a very broad and multidisciplinary hermeneutical approach. Natural phenomena are treated as socio-cultural facts in predominantly animistic settings where ancestor worship and adopted aspects of Hinduism prevail. The implementation of research findings refer ultimately to contexts. In essence, nothing can be easily transferred from one context to another, or be implemented in a context where the social, cultural and spiritual layout of a community is different from its original context. Any transfer, for instance, of environmentally context-bound plants or trees is a process of approximation to the circumstances of new surroundings and of negotiations between the 'sender' and the 'receiver' of two different communities. First of all, the frame of reference has to acknowledge that an adaptation or transfer of a tree species, or a management pattern, from one tribal context to another is a complex process that involves many people. Selecting and appropriating the items a community regards as useful for practical purposes can only happen if it is mediated through locally-operating agencies such as non-governmental organisations (NGO). Transfer and implementation of non-local knowledge, forest management patterns, the appropriation of non-timber forest products (NTFP), and modification of aspects of material culture, must seek local receiver-community's approval and acceptance which can never be taken for granted. A prerequisite for the acceptance of the process and the NGOs mediating role is its legitimacy and bargaining power in the local political setting. The traditional donor-initiated development project or programme perspective, however, will hardly be adequate and will have difficulties providing the legitimacy needed to promote the process of transfer of unfamiliar indigenous knowledge of plant and tree species, from the outside or by outsiders. The amalgamation and interconnectedness of ecological, economic and spiritual components of a particular socio-cultural context has to be worked out in a community. This is provided that there are problems which are believed to be successfully approached with the help of an intercultural transfer of knowledge.

Research helps to identify contexts that may be suitable for transferring local indigenous knowledge or techniques. The more similar or congruent these contexts are, the greater the chance for any kind of transfer with a positive response from the affected community. The adaptation of indigenous knowledge and technology through distribution is based on the assumption that potentials of 'traditional technology transfer' are there and have to be identified by research. As already stated above, the initiative for a transfer has to come from local communities themselves, or NGOs with good credibility and political influence at the local and regional level.

The process of social and cultural appropriation of non-local phenomena may lead to a sustainable intercultural adaptation of indigenous knowledge for forest use and management, if it is indeed a transfer that contributes to the wealth of the adapting community, and does not contradict cultural norms and values. The process of transfer of research findings from one 'forest world' to another is not predictable in its results. Ultimately, it depends on the responding community whether, or how, it will go through a process of appropriation. Sustainability can only be attained by self-induced efforts by the local population in co-operation with NGOs or local traditional institutions with a good public reputation among the tribal population. The task would be to promote local knowledge of forests and natural resource use among communities whose environments are prone to problems which could perhaps be solved by the transfer of indigenous knowledge.

A perceived lack of applicable traditional low-cost knowledge might be mediated through NGOs who while being involved in research, spread the message that indigenous knowledge can be a valuable source of economic security and cultural identity. The consciousness of having and using appropriate concepts of social resource use, based on a holistic approach towards one's environment, may in some cases be a greater incentive than to be integrated into the mainstream of a developing economy by a bureaucratic administration. Clients, who will depend more and more on a state bureaucracy cannot cope with degrading environments and social erosion in the long run. Once an administration has created or favoured clients who are used to demand subsidies, services and legal access to resources in exchange for promises to vote for certain political parties, these clients will become dependent on this paternalism and may cease to care for themselves. The implementation of research findings on indigenous knowledge, in contrast, is a socio-political process to achieve the strengthening of local self-reliance. In any case, such a process will contribute to the accountability of local village communities. In a phase of administrative deregulation, this could be an important step towards a culturally-appropriate development agenda to promote the self-empowerment of local communities.

Due to environmental and social changes, forest dwellers have become market-oriented to varying extents. Over the last decades, previously semi-nomadic shifting cultivators have adopted sedentary agriculture in the plains wherever possible, abandoning the hill forests. This is still an ongoing process linked to the tribal development programmes of the Government of Orissa, such as the Integrated Tribal Development Agency (ITDA) which has directed the lifestyles of the tribal forest dwellers into the mainstream thus disconnecting them from their traditional agricultural practices.

Forest-dwelling tribes within the regional and national political set-up

For an assessment of the forest dwellers' livelihood in the remoter areas of Orissa it is important to realise that forests are a highly aggregated amalgamation of economic, cultural and environmental factors. Those who live in the forest perceive its bounty for daily consumption and marketing opportunities for their self-reliant way of life. Outsiders who draw economic wealth from the same forests are generally urban people — almost all of them are non-tribals — who challenge the independent traditional lifestyle of the tribal population by depriving them of their natural resource base. The Orissa Forest Department and tribal development agencies as representatives of the State administration have the difficult task of mediating between the quest for natural resources to meet the needs of the non-tribal rural and urban population, and the forest dwellers. The tribal-non-tribal interface matters as far as the sustainable management of forests and agriculture is concerned. Land, and particularly land reserves, including wastelands and degraded forest land that can be converted into arable land is — particularly in a predominantly agrarian society — a highly political issue. Forest legislation and customary entitlements to use land, forests and forest products, often contradict each other and are matters of increasing litigation in areas with a predominantly tribal population. A mediation of economic and political interests between the tribal and non-tribal users is difficult for the Forest Department, whose task is to appropriate all forest-related goods on behalf of a state that has alienated the tribals from their freedom of usufructuary rights and has little legitimacy from their point of view. Commercial exploitation by wood-processing industries and paper mills is threatening forests all over India. An ever rising demand for timber and wood products, as well as NTFPs, has a high ranking priority in regional politics, as an important issue for local economies which often provokes corruption. At all levels of the rural agrarian society there is an increasing eagerness to gain access to forest land and forest resources which are claimed by forest-dwelling tribes for themselves. Consequently, it is important to distinguish between those sections of the society who live in and with the forest, and those who live from it (i.e. exploit it without being entirely dependent on it). It is therefore necessary to make a distinction between users of the entire range of products which are interwoven into the fabric of tribal forest life, and those who benefit commercially from an exploitation of select goods from the forests with a good market value.

The socio-economic distribution of local products for local well-being or for national wealth is reflected in the state's forest policy. At the local and regional levels, forest policy is covered by the Tribal Development Agencies which emphasise the economic situation of the tribal population and were established to provide fair prices at government fixed rates for their forest products (Fernandes *et al*. 1988). Until now Indian national forest policy of resource use mainly covers timber and seeds (*ibid*.:132). Minor forest produce (MFP) or, as it is called nowadays, NTFP, was always an economically relevant and expanding sector of commercial interest as, for instance, in the collection of *kendu* leaves for the tobacco industries, or the harvesting of bamboo for the paper mills.[1]

The government's political vision at most policy levels is to integrate the tribal population into the mainstream Indian civil society. In the case of forest dwellers, integration practically means the appropriation of their living space by the State and their alienation from their previous lifestyle, with compensation in the form of social services. Housing, health services and primary education are, amongst others, provisions to 'uplift' the tribal population, as the official jargon puts it. Development thus means the separation of those who are mainly using natural resources as primary users, in a way which is difficult to be controlled by the forest administration, from the deliberate use which was granted to them more or less willingly by

---

1. According to the 1981 census, the total revenue from logging in 1979-80 in Orissa was about 179 million IRs (all India: about 3,76 billion IRs); minor forest produce 88 million IRs (all India: about 1,36 billion IRs). *See* Govt. of India, 1987.

forest legislation since the first Indian Forest Law in 1865. It is important to see the findings of the 'Man and Forest' research network in the light of this political situation, and its strategic value as a scientific resource management capacity from a local perspective, as well as from a general political perspective of the future social development of India's rural areas.

In a period of administrative deregulation and an increasing financial crisis of the State, indigenous environmental management has an appeal for environmentalists as much as for economists and politicians. Local communities with an almost self-sustained livelihood, and a by and large sustainable management of the natural resources in their immediate surroundings, have a great environmental and political value. This sustainability is not to be understood in the sense of an active conceptualisation with an economic rationale of resource use, but in a very broad and highly flexible mode of subsistence. A rather low population density, and diversified patterns to use or to spare particular forest products from exploitation, does not put pressure on single food items or raw materials. This mode of 'exploitation at ease' can only happen if a forest-dwelling community has indigenous knowledge at its disposal that puts it in a position to survive in times of utmost scarcity. Economic competition among tribal groups, or between tribal and non-tribal groups, and the forest administration which imposes political restrictions on the use of forests, affects their lifestyles and prospects for future development. Local knowledge of semi-independent communities who know their environment very well is an important source of self-regulation and management at low costs for the state administration. In the tribal-non-tribal interface the importance of indigenous knowledge and its merits, as well as the problems of its limited scope of application, have to be taken into consideration. Most important in this respect is a reassessment of the opportunities as well as the limitations of integrating forest-related knowledge of the tribal population into a State or national forest policy.

Over the last two decades, scientific research has attributed a lot of prestige to indigenous knowledge attracting political attention and, and thus the self-esteem of a notoriously marginal section of Indian society is being raised. Another aspect is the encouragement to participate in decision-making processes at various levels, for instance in the process of JFM with the help and mediation of local and regional NGO. These processes can only succeed, however, if the political bargaining power of tribal populations is increased through research on the ecological potential of their knowledge. Natural resources conservation policy may increasingly make them partners to be considered more seriously in environmental planning processes.

Prospects for forest management in Orissa and India

From the experiences of this research project some generalisations can be made as far as the overall exposure of the forest world of tribal people to social change is concerned. Forest environments are under stress and are politically contested domains (Pathak 1994). Some of the points considered to be the most important social and political issues are listed here.

Those who think about opportunities and practical options to preserve forests have to be aware of the fact that the whole of the social structure of a community is related to the existence of these forests. Hardly any concept of appropriate forest management can afford to neglect a proper information base of scientifically-approved knowledge on what people who perceive themselves as indigenous know about forests. Economic self-reliance and political self-determination of resource use, as far as the degree of market integration is concerned, is a prerequisite to maintain a socio-ecological equilibrium to face the challenges of social change, and to meet the demands of the State administration. Local economic development is only sustainable if based on these prerequisites. Forests which are inhabited or used and managed by tribal populations who rely on their indigenous knowledge should be treated as stakeholders who try to maintain this equilibrium. The political positions of the tribal, the non-tribal population, and the forest administration are very distinct, but also partly overlapping and contradictory. The challenge for all of them is to come to terms of co-operation with one another. Their patterns of communication with each other with respect to forest policy matters will decide on the potential and actual relevance, as well as the feasibility, to negotiate solutions for practical problems.

The research findings of this project show that there is a link between local concepts of care for the forest world of a tribe as a whole, and an eco-friendly forest management from an environmental science's point of view. On the other hand, it is rather obvious that non-local people, without any attachment to the forest they are economically exploiting or administering, lack the cultural affection and intrinsic commitment to take care of the forest world which is characteristic of so many forest-dwelling communities. Poverty, economic marginality and opposition to programmes trying to involve them in social development activities have — although perhaps not always — had an ambivalent effect on forests. They may lead to both conservation as well as the destruction of forests and the depletion of forest resources. The most important guestion seems to be whether the social cohesion of a community makes it politically accountable as an organised social unit, and whether it is ready to co-operate with other village communities, NGOs and government authorities.

JFM has to become aware of the fact that forests can have many socio-cultural meanings for different groups and may serve many purposes in the development of a particular society. What political visions are implemented for the development of a rural area, and how the affected population perceives its future life is important for the survival of a local indigenous culture. As long as indigenous knowledge of forests is valued and considered to be an important representation of the cultures of forest-related communities, there will be international interest to document regulations to counter-balance scarcity or improper management. Once, the level of economic development is improved, it might reduce the pressure on forests as well as on forest-dwelling communities and their indigenous land use traditions.

In a tribal-non-tribal-State interface there are no immediate 'solutions' to forest-related problems in the strict sense of the term, only contributions to tackle the transitional inadequacies of using or managing forests. Regulations have to be balanced in the political arena between the directly-concerned population and the interests of other sections of a society using forest resources. Forests are very often burdened with the negative symptoms of societies in transition. A broad scope of investigation of the socio-cultural character of these symptoms is therefore essential. Interventions through afforestation programmes or social forestry projects are often unable to achieve solutions to forest-related problems, because they bypass fundamental problems and thus add to their complexity. Once resources have been created through afforestation, socially approved distribution patterns for the new wealth must be found. A question mark must also be placed on the sustainability of these 'technical' measures.

Research on indigenous knowledge of forests is essential because it is a prerequisite for the application of culturally diverse forest management strategies. Biodiversity preservation can only be achieved through cultural diversity and by supporting different forms of knowledge in as many tribal communities as possible. Knowledge of particular environments built up over centuries can only be provided by the experience of local populations.

A suitable approach to manage India's forests combines research activities with parallel social measures performed by local NGOs in relevant social sectors. Socio-political interactions between forest-dwelling tribes and the local non-tribal population, as well as the forest administration, have to be taken into account regarding their particular outlook on future developments. The legacy of India's persistent forest culture has embarked on a new phase in which its age-old traditions are challenged by social, political and environmental changes. The contribution of Indian civilisation to overcome the emerging forest crisis on the Indian subcontinent could be to remember her own traditional roots, and find ways out of the dilemma of economic development through the preservation of natural resources and cultural values.

Conclusion

The tribal life-world favours an overall spiritual outlook on environmental phenomena consolidated by social processes. This concept is basically not far from the Western understanding of environmentalism, before it became disenchanted during the era of the Enlightenment when space became the Cartesian *res extensa* and knowledge became *res cogitans*. Both space and knowledge became values in themselves and other values had to find their way back into modern society as morality or ethical codes of conduct, in order to make modern civilisation more humane, eco-friendly and less destructive. Nowadays, after environmental problems have found their place on the political agendas of the economically developed nations, lessons can be learned from those who have been taught to change their lifestyles in the West. The environment and development summit in Rio de Janeiro in 1992 has, among many other objectives, encouraged the investigation of traditional, local and indigenous development concepts in order to meet the social and environmental crises of the decades to come. What Indian tribal culture can contribute in this context ten years after the Rio summit, is perhaps the message that only a limited exposure to scarcity can result in the proper management of the affluence of forest wealth. It is a merger of utilitarianism and spirituality that ends up where it started from: the tribal world-view as an inseparable encompassing whole of nature and culture.

To rethink Indian forest culture in this sense points to the fact that largely independent forest-dwelling communities manage their habitats on their own, whereas the national forest policy administers resources on behalf of an anonymous and nation-wide commercial demand. The former have achieved a continuity to a remarkable extent where the latter will fail if it does not succeed in finding a way where space, knowledge and values amalgamate into a holistic concept of a coexistent life-fostering forest policy. Tribal world-views on their forest worlds thus represent an important aspect of continuity in Indian culture and civilisation that also has a message to convey to other civilisations.

Notes

Adhikary, A.K., 1991, 'The Tribal Worldview: Changing Perspective', In: Saraswati, B. (ed.), *Tribal Thought and Culture. Essays in Honour of Surajit Chandra Sinha*, New Delhi: Concept Publishing Company, pp. 105-20.

Banwari, 1992, *Pancavati. Indian Approach to Environment*, tr. A. Vohra, Delhi: Shri Vinayaka Publications.

Das, R.P., 1988, Das Wissen von der Lebensspanne der Bäume, *Surapalas Vrksayurveda*, Stuttgart/Wiesbaden: F. Steiner Verlag.

Elwin, V., 1954, *Tribal Myths of Orissa*, London: Oxford University Press.

Fernandes, W.G., Menon, P. Viegas, 1988, *Forests, Environment and Tribal Economy. Deforestation, Impoverishment and Marginalisation in Orissa, New Delhi*, Indian Social Institute.

Fernandes, W. (ed.), 1996, *Drafting a People's Forest Bill: The forest dweller – social activist alternative*, New Delhi: Indian Social Institute.

Gadgil, M., Guha, R., 1993, *This Fissured Land. An ecological History of India*, Delhi: Oxford University Press.

Government of India, 1987, Ministry of Planning, Dept. of Statistics, Central Statistical Organization, *Statistical Abstract India 1986*, New Delhi.

Pathak, A., 1994, *Contested Domains. The state, peasants and forests in contemporary India*, New Delhi: Sage Publications.

Poffenberger, M., McGean, B. (eds.), 1996, *Village Voices, Forest Choices – Joint Forest Management in India*, Delhi: Oxford University Press.

Prasad, M. (ed.), 1987, *Tribes – their environment and culture*, Delhi: Amar Prakashan.

Seeland, K., G.B. Patnaik, K.K. Patnaik, H.C. Das, M.K. Jena, P. Pathi, S.C. Behera, 1995, *Indigenous Knowledge On Forests. An Enquiry into the Worlds of Kuttia Kondhs and Saoras of Orissa, India*, Cuttack: Council of Cultural Growth and Cultural Relations (2 vols.).

Sinha, S.C., 1981, 'Tribes in Indian Civilization: transformation process in modern India', *Man in India,* 61 (2), pp. 105-42.

Sharma, B. D., 1995, *Globalisation – The Tribal Encounter*, New Delhi: Har-Anand Publications.

Shiva, V., (ed.), 1991, *Ecology and the Politics of Survival. Conflicts over natural resources in India*, New Delhi: Sage Publications.

3

Man-Nature Relationship of the Juang of Juangpirh
(Kendujhar District, Orissa)

Nityananda Patnaik

Introduction

THE biogeographic region of India comprises many ecosystems of different types. In India, with 2 per cent of the world's landmass, there are 7 per cent of the recorded species worldwide. Of the 45,000 plant species, about one-third are flowering plants and the others non-flowering like fungi, ferns, mosses and algae. Of the 81,000 animal species, nearly two-thirds of all amphibians (frogs, toads, newts) are found in India. Many indigenous beliefs, practices and prudent use of resources have helped to conserve these natural resources.

During the last century massive destruction of nature resulting in extensive and wide scale-depletion of natural resources began all over the country. The onslaught dealt a death blow to the traditional ideology and tribal practice of plant conservation resulting in depletion of natural resources. For instance, the Juang used to treat all elements of their landscape as their kin and regard themselves as part of the 'community of Being' which are jointly responsible in the making of the landscape of the Juangpirh or the Juang country. They used to attribute sacred qualities to them and therefore observe restrained use of resources. They have also a variety of practices which regulate the use of resources ensuring the long-term sustainability of resource use. Some examples of traditional practices have been given in the chapters that follow.

The fabric of socio-cultural life of the forest dwellers is interwoven with forests and with other constituent parts of nature including their management. A holistic study of the life and work of these people using the hermeneutic approach gives a vivid picture of this fabric. Such an ontological study of Being reveals what the community of beings of a particular ecosystem, say forests, animals, hills, streams, soil, water, and rock, means to the people and how this meaning is reflected in their cultural perceptions. The notions which these people have about natural resources have been developed out of these perceptions and their linkage with their day-to-day life. The main questions which require analysis are:

1. How are different kinds of natural resources (trees, forests, animals, mountains, streams, soil) perceived by the Juang?
2. How are they used and exploited?
3. How are indigenous knowledge and traditional practices applied in the management of conservation and for the long-term sustainability of resource use?
4. What processes of destruction and depletion of resources have been going on in the area, and with what consequences, and to what extent are the people aware of these processes?
5. What role do the people play to check destruction and what modes of restrained use of resources are practiced in response to the need for protection and proper management ?

Kendujhar district

Kendujhar was formerly one of the states of Orissa ruled by feudal chiefs. From AD 1128, until it was merged with the present State of Orissa in 1948 as a separate district, there have been 37 chiefs who held the *gaddi* (throne) one after another and ruled the State as independent feudal lords. Kendujhar district is situated in the northern part of Orissa and lies between 21° 1' and 22° 10' northern latitude and between 85° to 11' and 86° 22' eastern longitude. It is a landlocked district and its geographical area is 8303 sq km. It has 3 subdivisions, 7 tahsils and 13 blocks. The Juangpirh and Bhuinyapirh which cover the study area are situated in the hilly region of the Banspal block of Telkoi taḥsil in the Sadar subdivision of the district.

The district consists of 2067 villages of which Banspal block comprises 162 villages. The total population of the district is 1,337,026 as per the 1991 census. The Banspal block comprises 70 647 people out of which the scheduled tribe population is 79.29 per cent. Of the scheduled tribes inhabiting the district, the Juang and the Hill Bhuinya are the two most traditional and their population was 79,740 in the district (15,467 Juang and 64,273 Bhuinya population) as per the 1981 census. According to the information collected from the ITDA, in Kendujhar district there are 124 Juang villages and the Juang population is about 19,000 spread over 4,068 families.

STRUCTURE AND RELIEF

The main hill ranges of the district which run down the centre from the north-east act as the divide between the Brahmani river in the north-west, and the Baitarani river in south-east. Several spurs diverge from this main divide in different directions and make the study area hilly and rugged. The highest and best-known peaks are Gandhamardan (1061 m), Mankadanacha (1110 m) and Gopatganga (982 m). The hill slopes are generally concave in shape and the valley bottom is bowl-shaped rather than V-shaped. The hill-tops are generally flat and therefore, like the hill-slopes where shifting cultivation is practiced, the hill-tops are also used for raising crops. The valleys have very rich alluvial soil suitable for paddy cultivation. These valleys are intersected by numerous hill streams which during the rains become raging torrents, hurrying to discharge their water into the river Baitarani. Unlike the valleys, the hill-tops and the hill-slopes are covered with a thin layer of soil and therefore subject to erosion. Since the soil cover is not thick and deficient in plant nutrients, shifting cultivation is practiced in those places on a rotational basis with a fallow period for the sake of the recuperation of the soil. Laterite and red soil are common on the hill slopes. Sesqui oxide is the common ingredient of the soil whereas the Ferricoxide content is low. Other minerals present in small quantities are Potash, Phosphate and Nitrogen.

SURFACE DRAINAGE

The study area is confined to the catchment area of the river Baitarani whose source of origin is Gopatganga Pahada near Gonasika village. The river almost flows from south to north through the study area till it reaches the northern border of the district and from there turns to the south to come down to the plains flowing into the Bay of Bengal. With its numerous tributaries, the river causes floods during the rainy season. The study area is seriously prone to soil erosion and the river Baitarani carries away the eroded soil and deposits it in the plains of Orissa. Almost two-thirds of the study area is drained by the river. The State Government has recently built check dams at one or two places and the water that is checked is harvested for irrigating valley lands nearby.

MINERALS

Of the different mineral deposits the iron ore series of Gandhamardan hills are economically the most important. The ores are extracted from the hills at Suakati, the gateway to Juangpirh and Bhuinyapirh, and transported to Paradip where they are exported to Japan and other places. The mining activities at the Gandhamardan hills have absorbed many unskilled tribal labourers most of whom are Hill Bhuinyas of the locality.

FORESTS

The forests of the study area are of the monsoon type classified as tropical moist deciduous forests and tropical semi-evergreen forests. The *sal* forests with mixed species represent the climax vegetation whereas the dry deciduous forests and thorny xerophytic communities characterise various stages of degradation. Within these broad divisions of monsoon forests occur other characteristic vegetation types depending upon local habitat and ecological niche. The tropical moist deciduous forest is characterised chiefly by *Shorea*, *Terminalia, Diospyros* and *Schleichera* species. The characteristic associates are *Buchanania latifolia* and *Terminalia tomentosa* and the typical shrubs are *Wendlandia tinctoria* and *Woodfordia fruticosa*. Grasses are numerous, the most conspicuous being *Along* and *Sabai*. Bamboos were abundant in the past but have completely disappeared due to overexploitation.

GRAZING LANDS

The domesticated animals which are most common in the study area are cattle and goats. The statistics given by the Directorate of Economics and Statistics for Kendujhar District show that Banspal block has a forest cover of about 33,545 hectares. Other types of land use are miscellaneous tree crops and groves which cover 193 hectares. Permanent pasture and other grazing lands cover 1073 hectares and cultivable wasteland covers 3876 hectares in Banspal block. The net area sown is 9737 hectares. This relates to land at the valley bottom and not at the hill-slopes and hill-tops where shifting cultivation is practiced. The area under shifting cultivation is recorded to be 40638 hectares according to satellite imagery supplied by the Directorate of Soil Conservation.

ROAD COMMUNICATION

The National Highway No. 6 (Calcutta-Bombay Road) passes through Kendujhar town with Juangpirh on its left side and Bhuinyapirh on its right, and runs up to Sambalpur and beyond the State Highway No. 11 (Cuttack, Chaibasa, Ranchi Road) which passes through Kendujhar town. Two other important places in the tribal area are Gonasika, the centre of Juangpirh, and Banspal, the block headquarters, now connected by all-weather roads. The road which goes up to Gonasika starts from a village called Bayapandadhar located at a distance of 7 km from Suakati, both on the National Highway No. 6. The Gonasika-Bayapandadhar road is at a distance of 10 km. The road which goes up to Banspal block headquarters starts from Suakati village and is 13 km long. The hill streams which cut across these two village roads make them incommunicable during the rainy season when they are in spate.

CLIMATE AND RAINFALL

The climate of the study area was very pleasant when there was thick vegetation cover over the whole territory. With the depletion of forest the area suffers from oppressive heat during the summer season, but the winter is as cold as ever. The Juangpirh, having thicker vegetation, does not experience such oppressive heat as does the Bhuinyapirh. The hot season starts from the beginning of April when temperature rises rapidly. May is the hottest month in the year. The rainy season starts from the later part of the month of June and continues for 3 to 4 months. The area receives heavy rainfall in the months of July and August. Winter starts in November and lasts till February.

The Juang

The Juang are a tribe found nowhere else in India except in Orissa where they are in great concentration (nearly 95 per cent), with the traditional section living in Kendujhar district, and the acculturised section in Dhenkanal district. Gonasika, the heartland of the Juang territory (Juangpirh) is believed to be the cradle of the Juang, and it is from this point of origin that the Juang have dispersed to Dhenkanal. The Juang who live in Juangpirh or the Kendujhar Juang therefore designate themselves as *thanya* (local) and those of Dhenkanal as *bhagudia* (deserted).

Gonasika has become a famous pilgrim centre in the locality because the river Baitarani (the boat that sails with people in the ocean of this world to the other world or heaven) has come out from the nearby town, and after flowing towards the north and north-eastern part of the district descends to the plains finally flowing into the Bay of Bengal at Dhamra in Cuttack District. At Gonasika there is a Shiva temple built by the bank of the Baitarani river where a great festival called *Shivaratri* is celebrated and a congregation of pilgrims and devotees come to have a sacred dip in the Baitarani river. It is a great festive occasion in the area and both the tribals and the non-tribals take part in the worship, attracting tribals to Hinduism.

The Juang belong to the proto-Australoid racial stock and to the North Munda language group. The Juang language is a spoken language and has no script. Formerly, they used to wear a leaf apron which they have long since, abandoned but local non-tribal people still call them *patua* (leaf-wearing people).

The settlements of the Juang tribe which are scattered in the Juangpirh have no regular village street and the houses are very small. At the centre of a settlement the village goddess Gainseri is placed under a plant. Beside this, the bachelors dormitory (*mandaghar*) is built which is the biggest hut in the village. It has walls on three sides and an open front. On the front side is a dancing ground the size of a badminton court. The elders of the village sit in the open front side and watch the boys and girls dancing together at night.

The *mandaghar* is a kind of club house where the fire is kept lit day and night throughout the year and the bachelors sleep at night around the fire place. The elders gather here in leisure hours and spend time gossiping or discussing all kinds of village affairs. The site for swidden cultivation, and the dates for observing rituals and other socio-economic activities are fixed in consultation with one another in the *mandaghar*. Seeds of various crops are stored here and those sanctified for sowing are distributed among the residents of the village in front of Gainseri. Visitors to the village are entertained at this place. It is kept under good repair by contributions of community labour.

A Juang village was formerly composed of households of a single lineage belonging to a single clan (*bak*). However, it is now a multi-clan structure. Thus the village is exogamous. There are two categories of villages: *Kutumba* villages (those inhabited by families of same lineage) and *Bandhu* villages (those inhabited by families with which marriage can be held categorisation of settlements is ecologically determined. Among the Juang, lands under shifting cultivation are held in common and cultivation, which is extremely hazardous, requires reciprocal help. These conditions are easily fulfilled when a village is composed of agnates who voluntarily help each other in clearing the forest and in other activities connected with shifting cultivation.

Like in their dormitory organisation, the Juang have a system of age-grades both among males and females. The age-grades among the males start from childhood to old age. When a boy attains the age of 15 years he becomes a *kanger,* eligible to become a full-fledged member of Majang after undergoing certain rituals. When he is married he ceases to be a member. In the case of a girl who attains puberty at the age 13-15 years she becomes a *selan* and is eligible to be a member of the dormitory after undergoing certain rituals. When she gets married her membership is removed. The persons of different age grades have specific functions to perform in the village in general and specific roles to play in the dormitory in particular.

LIVELIHOOD

The Juang were formerly hunters and gatherers, and nomadic like the Birhors. But as time passed they took up shifting cultivation and settled down in permanent settlements. Shifting cultivation continues to be their main source of livelihood. Further, they have taken up wet cultivation in valley lands. Starting from the kitchen garden and progressing through the uplands near villages up to the hill-slopes, and in some places up to the hill-tops, they grow a great variety of crops. (Annexure III). Most of these crops are grown in swiddenland. Some of them, specially vegetables and condiments, are also cultivated in kitchen gardens and in fertile spots near village sites. Their indigenous knowledge guides them in crop husbandry. They know the soil-water-sunlight requirements of each crop as well as seasonal conditions. They also know the limits of their lands, the kind of soil and the climate of the area.

The Juang use a simple digging stick and hand hoe to till the fields. They also have remarkable dances imitating the movements of animals and birds, as well as legends depicting their origin from trees.

SHIFTING CULTIVATION

Shifting cultivation called *ekan* is the main source of livelihood of the Juang. A patch of swidden is cultivated for 3 years. The first year cultivation is called *surilok* and pulses of different types are grown. The second year cultivation is called *bualok* and the principal crop grown is paddy alongside different kinds of millets and cereals. The third year cultivation is called *nala* and *ramtila* as is grown in the first year. Mixed cropping is the rule but the cropping pattern varies. Some crops are grown along the border while others are sown in the middle of the plots. Some crops are not mixed with other crops and are grown in specific places, whilst others are grown side-by-side so that one supports the other. In mixed cropping, one advantage is that the crops mature at different times and therefore the harvesting is sequential. When one crop is exhausted, another crop is ready for consumption.

There are also other advantages of multiple cropping. There is less danger of crop failure. It provides a varied diet which in turn meets individual taste and digestibility. Different crops meet different needs as, for instance, the sale value of some crops is more than consumption needs; some crops take less time in cooking; some take more time requiring more fuelwood; some mature earlier; and others provide food during the lean period. Different crops have different needs and multiple cropping gives a higher overall productivity than monocropping. After the third year the swidden patch is left fallow for 5 to 7 years to recuperate. The practice of shifting cultivation is based entirely on traditional knowledge. The preservation of seeds, the selection of crops, cropping patterns, pest control, the storage of crops, and many other associated activities are based on indigenous knowledge.

Shifting cultivation is no longer providing a livelihood as it once did. Population has increased and there is no scope for expansion. The fallow period has shrunk from 15-20 years to 5-7 years now. The pressure of population has increased many times and the carrying capacity of the land under shifting cultivation is under strain. Even the valley lands which are now cultivated for growing paddy are in short supply and the simple agricultural practices do not help to get a better yield from the plain lands. Moreover, the forest cover in the area has shrunk considerably for various reasons including fuelwood collection; cattle and goat browsing; indiscriminate tree felling; the denial of rights to the local people leading to organised felling; the disruption of an in-built traditional concepts of carrying capacity; leading to a perception of forests as an open access resource representing a source of profit mainly for outsiders.

The consequences of deforestation are many. But the most serious ones are environmental degradation which threatens to countless species of forest flora and fauna including medicinal herbs and food plants affecting the health, nutrition and income. Central sources of subsistence include: fruits, nuts, seeds (*mahua* spirit and oil seeds), leaves, roots, shoots, animals, birds, insects, fish, and income by selling forest produce, construction wood and animal fodder. Now that there is a colossal depletion and heavy drain of natural resources life of the forest-dwelling Juang is threatened. A very large percentage (about 59 per cent) of earnings of the tribals comes from minor forest produce (MFP). The large-scale deforestation has made the tribals poor.

Statistics of MFP for the study area are not easily available however. Since the coverage under forest is greater in Bansapal block, particularly for Juangpirh, the figures indicate the relative richness of the study area regard to with MFP. A list of the MFP and the quantity collected in 1993 is given below:

1.	Kendu leaf	22593 Qtl
2.	Tamarind	6000 Qtl

3.	Hill broom	159356 pieces
4.	Myrobalan	7397 Qtl
5.	Gum	200 Qtl
6.	Nox vomica	200 Qtl
7.	Siali leaves	2500 pieces
8.	Sal seeds	41977 Qtl
9.	Mahua	4630 Qtl

[There are many other items which are available in small quantities, see Chapter 4.]

Indigenous knowledge system and practices (IKSP)

The term indigenous knowledge (IK) was used for the first time in a publication by Robert Chambers, in 1979 and in a book co-edited by Brokensha, Warren and Werner (1980).

> Klaus Seeland has defined it by saying that:
>
> Indigenous (Knowledge) means that something is originating locally and performed by a community or society in this specific place. It emerges as people's perceptions and experience in an environment at a given time and is a continuous process of observation and interpretation in relation to the locally-acknowledged everyday rationalities and trans-cendental powers. (2000:34)
>
> Knowledge is to be called indigenous, if it is bound to local experiences and takes its local world perhaps not as the only the existing, but as the most relevant of all, it is an authentic appropriation of Being. (2000:35)

In other words, indigenous knowledge is location and culture-specific knowledge. Being influenced by global or Western knowledge, one tends to forget that over the centuries human beings have been producing knowledge to live a balanced relationship with their natural and social environment in order to survive. Indigenous knowledge is specific to a given culture, location or society. It refers to a large body of accumulated knowledge with which the people are able to manage their natural resources in order to subsist on a long-term basis. These indigenous practices of natural resource management are applicable to agriculture, forestry, animal husbandry, human and animal health, soil conservation, water harvesting, cottage industries and many other sectors. Indigenous knowledge is handed down from generation to generation through observation, practice and enculturation. Indigenous knowledge is an information bank which supports the communities in adapting themselves to changing circumstances.

The role of IKSP in participatory and sustainable approaches to development is important. The lessons and principles which have emerged out of the short span of the study among the Juang are listed below:

1. IK is overlooked by planners and executives alike.
2. IK has both strengths and weaknesses from both emic and etic points of view.
3. IK can be compared and contrasted with global knowledge.
4. IK is characterised by both continuity and change.
5. IK case studies may be collected and used by development practitioners, research scholars and voluntary workers.
6. IK is a cost-effective component in the development process.
7. IK and Global Knowledge (GK) systems are complementary — each benefitting from the other (i.e. mixed cropping *vs.* monocropping).
8. Capacity-building is achieved by working with IK and through International Organisations (IO).
9. IK reflects ways of identification, categorisation and classification of plant species and animal species.
10. Some IK is common knowledge known to all members of the community. Some IK is highly specialised and confined to specialists.

In regard to recording IK, the technique of participatory Rapid Rural Appraisal is followed. The researcher or investigator transects the terrain with local people, asks them to give the names of the land forms and types of vegetation. Different land forms are often perceived by the local people as having different problems, uses, and characteristics. The community in question uses its language to identify, categorise and classify objects of nature according to a taxonomic principle. Certain domains of knowledge where taxonomic structure applies less include such systems as volume, time and distance. The researcher also finds that there are certain features of resource management which are associated with supernatural sanctions, taboo and other aspects of the Juang cosmological system. The recording of the knowledge system should take into account both inside and outside perspectives, so that it is possible to compare and contrast IK with global knowledge.

PRUDENT USE OF NATURAL RESOURCES

An observation of the life of the Juang shows that they depend upon their habitat to meet most of their subsistence needs. They are still hunting and gathering; trapping birds and other small animals; fishing in streams; cultivating hill-slopes, hill-tops and valley lands. They are raising fruit trees, keep livestock, they are weaving mats, carving, wood collecting fuelwood for domestic cooking, using herbal medicine and animal products for preventive and curative purposes. Above all, they are still staging their life drama and match-making, composing songs, and performing magico-religious rites and rituals on the sylvan background. Truly the Juang are a good example of a people depending almost exclusively on local natural resources for their living.

Unlike a commercial group with access to resources from an extensive area sustainability the Juang, whose resource consumption is low, are bound to be prudent in their use of the local resources. In the case of commercial users they generally have the option of moving on to another locality on exhaustion of the resources in a given area. They tend to focus at any time on the resources that bring them the greatest profits. One example of such commercial users of biomass resources is the paper and plywood industries. As soon as the raw materials in the locality close to the industries are depleted, they move on. They also have options to switch to a different type of resource, for instance to Eucalyptus if bamboo is exhausted.

In the case of the Juang a different attitude towards local resources is noticed. Firstly, the size of the resource base on which they depend to meet their subsistence needs is very small and they are unlikely to leave their homeland so easily. Moreover, unlike the commercial exploiters who are under the clutches of administrators and their political masters the Juang are free. The rear total dependence of the Juang on locally collected resources continues to this day. Since women are specially involved in the gathering of such biomass they are particularly concerned with the maintenance of such materials.

Things have changed as a result of regulatory policies and practices. The tribal communities have lost control over their local resources. Such control has been taken over by the State through its Forest, Revenue and other departments. With the State taking control, tribal communities no longer enjoy any right to control the use of common property resources by their own members or through their own institutions. The local tribals are anxious that the fruits or their own prudent use will be exploited by somebody else and are therefore losing their motivation to exercise a careful use of resources.

TRIBAL KNOWLEDGE OF PLANT CONSERVATION

The wisdom of tribal people and their imposition of taboos, have helped a great deal in plant conservation. Some measures of conservation are mentioned below.

Conservation of plants associated with totems

Totemism and fetishism are essential aspects of tribal religion. The plants associated with animism are worshipped and not destroyed. Similarly, taboos are observed restricting the use of plant parts to certain seasons and for the purpose of worship or propitiation of supernatural powers. Certain landmarks like a very big tree, or a patch of vegetation with some specific plants are believed to be the dwelling places of ancestral spirits and supernatural powers, and therefore the plants of those places are not damaged. In spite of some ecological and environmental changes, the ideas about plants connected with totem, taboo and clan have not changed.

Conservation by restriction

Certain plants near village sites are protected for specific purposes. If any parts of these plants are needed, the reproductive parts are left for regeneration. Certain medicinal plants are not harvested on certain days and at certain hours of a day. Many such restrictions play a vital role in the conservation of genetic resources.

Imposing restrictions on cutting and peeling plants

The tribal people avoid excessive peeling of the bark from any single plant. They peel bark fibre from the branches of *Bauhinia Vahlii* thereby protecting the main trunk of the plant. They never cut fruit bearing plants like *mahua* (*Madhuca longifolia*) as its flowers, fruits, and seeds serve as food during the lean months. There are restrictions on cutting trees during the post-natal period because the belief is that by doing so, the baby will fall sick.

Imposing restrictions on premature plucking

Tribal people do not use a medicinal plant before its flowering and fruiting. They rarely uproot the whole plant for the roots, tubers, rhizomes, which are of medicinal value to them. If they have to do so under certain circumstances, they leave some roots and tubers at the place from where the uprooted the plant can regenerate.

Restrictions on lopping, tapping and plucking

The leaves of certain plants like Asan (*Terminalia tomentosa*), are generally used as fodder. The tribal people are very careful collectors of fodder so that the tree is not hurt. Collection of gum, resin, latex and sap is not done from younger plants.

PRESERVATION THROUGH WORSHIP AND MAGICO-RELIGIOUS BELIEF

Many plants like Karam (*Ardina cordifolia*), *sal* (*Shorea robusta*) and *Garia* (*Curcuma* sp.) are worshipped. The tribal people are very careful not to cause any damage to these plants. The Juang know of a vast number of uses of many plants (trees, herbs, shrubs, climbers). One such useful plant is a grass called *along* which is used for thatching because of its durability and its quality of not being eaten by white ants and rats. The Juang have been extracting services from nature without large inputs since time beyond memory. Their practices have therefore been moulded to work closely with Nature.

Ethnobotany

Ethnobotany deals with the direct relationship of plants with man, Which can be broadly classified into two groups, namely: (1) abstract, and (2) concrete. The abstract relationship of man with plants includes beliefs concerning the good qualities or bad effects of plants. It also deals with taboo, avoidances, worship, folklore and myths concerning plants. The concrete relationship includes the material use of plants for food and drink, masticatory use, medicine, house-building, domestic uses, agricultural tools, and plants in art, painting, carving, and other such uses.

The benefits derived from vegetation are numerous. A few such benefits are listed below:

1. Preservation of vegetation as sacred groves or through various taboos.
2. Conservation of germ plasm through cultivation of land races.
3. Regularity of agricultural operations through mythological associations with celestial bodies, seasons, and so on.

METHODS OF ETHNOBOTANICAL RESEARCH

The sources of ethnobotanical data can be classified into 4 broad categories. These are:

1. Fieldwork among the tribes and recording folklore about plants.
2. Study of secondary literature concerning traditional medicine, ethno-biology, ethnology, archaeology, and historical records preserved in archives and museums.
3. Study of records preserved in herbaria for information regarding the uses of plants.
4. Study of archaeological remains for evidences of past relationship of people with plants.

A few examples are given below by analysing the myths and folklore of the Juang.

Origin of natural objects

According to the Juang, in the remotest past Mother Earth was very unsteady and there were no human beings around. Abbreviating the myth, the *rishi*'s eldest son was killed and his blood sprinkled over the surface of the earth. As a result, the earth became steady and from the palms and feet of *rishi*'s son hills and valleys emerged.

Rivers and streams flowed from his nostrils and eyes and forests grew from his hair. Thus the earth became suitable for habitation.

Origin of clans

Rusi and Rusiani had 12 daughters and 12 sons. These 12 brothers lied to their father and the consequence was that they were swept away by the currents of water which gushed out in a great flood from the nostrils of the cow that was killed by them. In order to save themselves, each of the brothers caught hold of the objects which were present along the way in the flood water. Each of the brothers became father of the things that they caught hold of and therefore became kin. The names of the things which they got hold of are listed below:

Table 1

	Objects	**Clans**
1st brother	*Kirri tree*	*Kirriyari*
2nd brother	*Stone*	*Tangorpariya*
3rd brother	*Sarai*	*Sariya*
4th brother	*Champa*	*Champai*
5th brother	*Machan*	*Rundali*
6th brother	*Jetua creeper*	*Jetua*
7th brother	*Tortoise*	*Koicho*
8th brother	*Sarangi fish*	*Sarambi*
9th brother	*Lanjhia grass*	*Lanjhia*
10th brother	*Sidri creeper*	*Sariyari*
11th brother	*Salar tree*	*Salar*
12th brother	*Saiko grass*	*Saiko*

These relationships with plants and animals have led them to think that objects of Nature are their kin. The Juang depend on wood for fuelwood consumption. An estimate of the use of fuelwood by the people of Guptaganga is as follows. Daily fuelwood consumption per family comes to approximately 12 kg and annual consumption is 4380 kg. The village comprises 50 households and their fuelwood requirement per year comes to approximately 290,000 kg. The forest cover within the Guptaganga boundary is about 900 ha, thus the forest per household is 18 ha. According to this calculation, the yield of fuelwood per ha of forest comes to 243 kg and households show no extravagance in the use of fuelwood.

Indigenous methods of classification

Ethnobiology in the broadest sense of the term is devoted to the study of the complex set of relationships of plants, animals and human society. Broadly two basic questions are raised: (1) how and in what ways do human societies use Nature, and (2) how and in what ways do human societies view Nature? The first question is an economic one and the second is a cognitive one.

Ethnobiologists recognise the value of both economic and cognitive aspects and ask why human societies classify different natural phenomena such as soil, hills, flora and fauna, in the way they do. The field of ethnobiology dwells more on classification of natural phenomena rather than on economic and cognitive factors. Lévi-Strauss (1966), on the other hand, has argued that primitive people's familiarity with their biological environment can scarcely be of much practical effect. The paragraphs which follow give the classificatory system in respect of differentiation of species as determined by the Juang which was studied in the first phase of the research project in the Juangpirh of Kendujhar district.

DIFFERENTIATION OF SPECIES AS DETERMINED BY THE JUANG

The Juang are capable of distinguishing various plant species, a result of continuous exposure to the forest and a keen observation of plants. However, it is difficult for them to assign higher order categories. (see Table 2). In rare instances, they are able to distinguish between allied species and give names accordingly.

Table 2

	Juang Name	*Botanical Name*	*Characterstic features among species as distinguished by the Juang*
1.	**Saraka**	Sida acuta (*Malvaceae*)	Rigid stem, leaves lanceolate
	Bajranali	Sida cordata (*Valvaceae*)	Stem prostrate, leaves ovate
	Biskapura	*Sida cordifolia* (*Malvaceae*)	Stem densely hairy, leaves ovatecordate
2.	**Kalia Jahlai**	Diospyros lylvatica (*Malvaceae*)	Leaves elliptic

	Teren	Diospyros melanoxylon (*Ebenaceae*)	Leaves broadly ovate. Fruit size larger, edible
	Kalikendy	Diospyros malabarica (*Ebenaceae*)	Usually occurs near stream; leaves oblong; fruit size similar to *Teren* but covered with reddish hair. Fruit rarely edible
3.	**Ala**	Phoenix acaulis (*Arecaceae*)	Shrubs, leaflets, narrower, fruits small
	Khijiri	Phoenix sylvestris (*Apocynaceae*)	Trees, wide leaflets, fruits large
4.	**Kiring**	Holarrhaena antidysenterica (*Apocynaceae*)	Fruit thin
	Boi Kiring (Boi means larger size of allied species)	Wrightia arborea (*Apocyanaceae*)	Fruit thick, plant comparatively tall
5.	**Kaunria**	Hibiscus cannabinus (*Malvaceae*)	Stem reddish, flowers small
	Banakaunria	Abelmoschus crinitus (*Malvaceae*)	Stem green, flowers large
	Jamu	Syzygium cumini (*Myrtaceae*)	Leaves eliptic, fruits edible
	Koi Jamulia	Syzygium roxburghianum (*Myztaceae*)	Leaves wide
	Poi Jamulia	Syzygium cerasoides (*Lauraceae*)	Leaves ovate
6.	**Kala Khandakhai**	Litsea glutinosa (*Lauraceae*)	The inner bark
	Nali Khandakhai	Phoebe lanceolata (*Caesalpiniaceae*)	The inner bark is reddish brown
7.	**Bad chakunda**	Cassia occidentalis (*Caesalpiniaceae*)	Tall. Leaves with acute tip
	Chakunda (Probably, this name Chakunda has been adopted from another community in the locality)	Cassia tora (*Caesalpiniaceae*)	Short. Leaves with round tip

The Juang way of naming plants coincides mostly with the biologist's system of classification. However, in certain cases the Juang are not able to differentiate the species. Though plants have a distinct shape, the tribals sometimes consider a single feature to identify some specimens. Thus, the tribe uses a common name for different plants. This can be seen from the following examples.

According to the tribe, *Kiring* and *Boi Kiring* are two allied species. The latter is larger, in height and has thicker fruits. But, in most of cases, the tribe does not recognise a higher order, genus. Taking some common characters, the botanist puts the species under a genus and similar genus groups are put under a family. The Juang have different names for different species under a genus. For example, the name *Saraka bajranali*, and *Biskapura* are given to three different plants by the Juang and they know that these three plants are very alike in outward appearance. Through identification by a botanist it is revealed that *Saraka*, *Biskapura* and *Bajranali* are three different species belonging to the genus *Sida* under the family *Malvaceae*. The Juang identify different species belonging to the genus *Sida*. Similarly, two allied species, namely *Kala khandakhai* and *Nali khandakhai,* are distinguished by the colour of their bark. Thus *khandakhai* may be synonymous with the level of genus, but these two species belong to two different botanic genera. From this observation it can be seen that the Juang have their own rationale of plant classification.

The following examples are to elucidate the Juang concepts of distinguishing plant species.

1. *Madang*: Madang means parasitic plants growing on other plants, namely *Dendrophthoe falcata*, *Loranthus scrrula*.
2. *Sunsunia*: This name is given to two different plants. One is *Marsilea minuts* which belongs to the non-flowering plant family *Marsileaceae*. The other one is *Oxalis corniculata* which belongs to the angiospermic family *Oxalidaceae*. It is noteworthy that, although both plants are given the same name on the basis of the shape of the leaf, the Juang are aware of other distinguishable features. They know that *Marsilea* sp. occurs in swampy places and has larger leaves. On the other hand, *Oxalis* sp. occurs in moderately dry places, and has smaller leaves which taste sour. Both species are edible.

3. *Jabaka*: Jabaka means sticking. The fruits of certain plants stick onto the hair of animals. The plants with this feature are as follows:

    (i) *Bidens bilos* (*Asteraceae*)

    (ii) *Chrysopogon aciculatus* (*Poaceae*)

    (iii) *Cosmos caudatus* (*Asteraceae*)

    (iv) *Xanthium indicum* (*Asteraceae*)

4. *Kanbaluk*: The following herbaceous plants with small, round and hooked fruits are called *Kanbaluk*:

    (i) *Triumfetta pentandra* (*Tiliaceae*)

    (ii) *Triumfetta pilosa* (*Tiliaceae*)

    (iii) *Urena lebata* (*Malvaceae*).

Conclusion

The Juang depend primarily on shifting cultivation for their livelihood and also devote considerable time to food gathering and hunting in the forest. For various reasons, there has been extensive shrinkage in forest cover and forest resources. The practice of shifting cultivation has been very seriously affected. Two decades ago the recuperative cycle was a period of almost 20 years, but it has shrunk to 5-7 years due to increasing population and reduced forest cover in the area.

In spite of the depletion of natural resources, the Juang continue to live exclusively on local resources, particularly forest products, for their sustenance. The forest continues to be perceived as a renewable resource and part of their life support system. The commercial attitude towards the forest as a raw material to be exploited for profit without any consideration for the regeneration of vegetation is absent among the Juang. The Juang collect edible tubers, leaves, shoots, fruits, flowers, seeds, grains from the forest. They meet their requirements of fibre, timber, oilseed, gums, mat-making materials, fuelwood, fodder, herbal medicine and many other resources from the forest. Plants and animals are still a part of important parts of their religion, mythology, songs, festivals, totems, proverbs, metaphors, riddles, paintings and decorations.

As a result of their close contact with Nature their knowledge about natural phenomena is tremendous. They are able to predict when rain will fall and what crops are to be grown in which season. They are very knowledgeable about the role which, mineral salts in the soil play in the sustained growth of the plants. They know about the fertility of their lands, the kind of soil they have in the area, and their grasp of climatic and seasonal changes is impressive. It seems that the perception which helps them to perceive and bring an order into the plant and animal kingdom is based primarily on utilitarian factors.

The study of the principles of classification leads to a detailed ethnobiological investigation of a particular ecosystem. The question is what significance do their lifestyles represent in the industrialised world. In this connection, it should be remembered that in spite of tremendous advancements in technology, the forest-dwelling tribal communities are primarily dependent on plants and animals for food, medicine and other essential necessities of life. Ethnobiological investigation reveals the experience and expertise, knowledge and wisdom, of diverse ecological niches harnessed by generations of tribals. This information is of great utility in matters of a self-reliant food supply, disease control, and mitigation of environmental catastrophes.

Annex I

Medicinal Plants of Juangpirh

Menispermaceae

1. CISSAMPELOS PAREIRA Linn. — Climber

Papaveraceae

2. ARGEMONE MEXICANA Linn — Shrub

Capparidaceae

3. CRATEVA NURVALA Buch. — Tree

Dipterocarpaceae

4. SHOREA ROBUSTA Gaerth. f. — Tree
5. HIBISCUS LAMPAS Cav. — Tree

Sterculiaceae		
6.	HELICTERES ISORA Linn.	Shrub
7.	STERCULIA URENS-Roxb.	Tree
8.	AEGLE MARMELOS Corr.	Tree
Sapindaceae		
9.	SCHLEICHERA OLEOSA (Lour.) Oken.	Tree
Anacardiaceae		
10.	BUCHANANIA LANZAN Spr.	Tree
11.	MANGIFERA INDICA Linn. f.	Tree
12.	SEMECARPUS ANACARDIUM Linn. f.	Tree
Papilionaceae		
13.	MUCUNA PRURITA Hook. f.	Twiner
14.	PONGAMIA PINNATA (Linn) Pierre	Tree
15.	TEPHROSIA PURPUREA Pers.	Shrub
Caesaliniaceae		
16.	CASSIA TORA Linn.	Shrub
Combretaceae		
17.	COMBRETUM DECANDRUM Roxb.	Shrub
18.	TERMINALIA BELLIRICA (Gaertn.) Roxb.	Tree
Leeythidaceae		
19.	CAREYA ARBOREA Roxb.	Tree
Lythraceae		
20.	WOODFORDIA FRUTICOSA (Linn.) Kurz.	Shrub
Samyadaceae		
21.	CASEARIA ELLIPTICA Willd.	Small tree
Cucurbitaceae		
22.	BRYONOPSIS LACINIOSA (Linn.) Naud.	Climber

Rubiaceae

23.	GARDENIA LATIFOLIA Ait	Tree

Compositae

24.	ELEPHANTOPUS SCABER Linn.	Herb
	SPHAERANTHUS INDICUS Linn.	Herb

Plumbaginaceae

26.	PLUMBAGO ZEYLANICA Linn.	Shrub

Oleaceae

27.	NYCTANTHES ARBOR-TRISTIS Linn.	Shrub or small tree

Asclepiadaceae

28.	HEMIDESMUS INDICUS (Linn). Schult.	Creeper

Apocynaceae

29.	HOLARRHENA ANTIDYSENTERICA (Linn.)	Shrub or tree

Loganiaceae

30.	STRYCHNOS POTATORUM Linn.	Tree

Scrophulariaceae

31.	SCOPARIA DULCIS Linn.	Herb

Bignoniaceae

32.	OROXYLUM INDICUM (Linn.) Vent.	Tree

Acanthaceae

33.	ADHATODA VASICA Ness.	Shrub
34.	ANDROGRAPHIS PANICULATA Ness.	Shrub
35.	HYGROPHILA AURICULATA (Schum.)	Heine Shrub

Labiatae

36.	LEONOTIS NEPETAEFOLIA R.	Br.-Herb
37.	LEUCAS ASPERA Spr.	Herb

Amaranthaceae

38.	ACHYRANTHES ASPERA Linn.	Shrub

Polygonaceae

39.	POLYGONUM PLEBEJUM R. Br.	Herb

Loranthaceae

40.	DENDROPHTHOE FALCATA (Linn.f.) Etting.	Parasite

Euphorbiaceae

41.	EUPHORBIA HIRTA Linn.	Shrub
42.	JATROPHA GOSSYPIFOLIA Linn.	Shrub
43.	RICINUS COMMUNIS Linn.	Tree

Scitamineae

44.	COSTUS SPECIOSUS Sm.	Stout herb

Amaryllidaceae

45.	CURCULIGO ORCHOIDES Gaertn.	Herb

Liliaceae

46.	ASPARAGUS RACEMOSUS Willd. var. JAVANICUS Baker.	Prickly Climber

Gramineae

47.	HETEROPOGON CONTORTUS (Linn.)	Grass

Annex II

Wild Food Plants of the Juang of Juangpirh, Kendujhar District, Orissa

1.	AMARANTHUS SPINOSUS	*L. Amaranthaceae*	Tree
2.	BAUHINIA PURPUREA	*L. Caesalpiniaceae*	Tree
3.	BAUHINIA VAHLII. Wight and Arnott	*Caesalpiniaceae*	Robust climber
4.	BUCHANANIA LANZAN Spr.	*Anacardiaceae*	Tree
5.	CARYOTA URENS L.	*Palmae*	Tree
6.	CELOSIA ARGENTEA L.	*Amaranthaceae*	Tree

7.	COIX LACRYMA -JOBI L.	*Gramineae*	Tall grass
8.	DIOSCOREA BULBIFERA L.	*Dioscoreaceae*	Climber
9.	DIOSCOREA HISPIDA Dennst.	*Dioscoreaceae*	Climber
10.	DIOSCOREA OPPOSITIFOLIA L.	*Dioscoreaceae*	Climber
11.	DIOSCOREA PENTAPHYLLA L.	*Dioscoreaceae*	Climber
12.	DIOSPYROS MELANOXYLON Roxb.	*Ebenaceae*	Tree
13.	EMBLICA OFFICINALIS Gaertn.	*Euphorbiaceae*	Tree
14.	INDIGOFERA PULCHELLA Roxb.	*Papilionaceae*	Shrub
15.	IPOMEA AQUATICA Forsk. syn. I.reptans (L.) Poir.	*Convolvulaceae*	Creeper
16.	MADHUCA INDICA Gmel.	*Sapotaceae*	Tree
17.	MANGIFERA INDICA L.	*Anacardiaceae*	Tree
18.	POLYGONUM PLEBEIUM R.Br.	*Polygonaceae*	Prostrate herb
19.	SCHLEICHERA OLEOSA (Lour.) Oken.	*Sapindaceae*	Tree
20.	SHOREA ROBUSTA Gaertn. f.	*Dipterocarpaceae*	Tree
21.	SPONDIAS PINNATA (L.f)	*Kurz. Anacardiaceae*	Tree
22.	STERCULIA URENS Roxb.	*Stereuliaceae*	Tree
23.	SYYGIUM CUMINI (L.) Skeels.	*Myrtaceae*	Tree
24.	TAMARINDUS INDICA L.	*Caesalpiniaceae*	Tree
25.	WOODFORDIA FRUTICOSA (L.) Kurz.	*Lythraceae*	Tall shrub
26.	ZIZYPHUS OENOPLIA Mill.	*Rhamnaceae*	Climbing shrub
27.	ZIZYPHUS MAURITIANA	*Lamk. Rhamnaceae*	Shrub or tree
28.	ZIZYPHUS RUGOSA Lamk.	*Rhamnaceae*	Shrub or tree

Annex III

Cultivated Food Plants

CROPS

Oryza sativa (Rice - 15 Varieties)

Triticum spp. (Wheat)

Zea mays Linn. (Maize)

Sorghum vulgare (Sorghum)

Eleusine coracana Gaertn. (Finger millet)

Panicum miliare Linn. (Little millet)

Panicum miliaceum Linn. (Common millet)

Setaria italica Beauv. (Foxtail millet)

Pennisetum typhoides Stapf and Hubb. (Pearl millet)

Panicum crusgalli (Common millet)

Amaranthus sp. (Pseudo cereal)

PULSES

Cajanus cajan sp. Mill. (Red gram)

Dolichos lablab Linn. (Bean)

Phaseolus mungo (Black gram)

Vigna unguiculata (Green gram)

Vigna umbellata (Pulse)

Macrotyloma uniflorum (horse gram)

Mucuna sp. (Common cowitch/Cowhage)

STARCHES

Ipomoea batatus Lam. (Sweet potato)

Dioscorea spp. (Yams; more than a dozen varieties).

OILS AND FATS

Ricinus communis Linn. (Castor)

Brassica campestris (Mustard)

Guizotia abyssinica Cass. (Nizer)

FIBRE

Kaunria (Deccan hemp, Hibiscus cannabinusL.)

CONDIMENT

Pogostemon sp. (Aromatic spice)

Capsicum annuum Linn. (Chilli)

Zingiber officinale (Ginger)

Curcuma longa (Turmeric)

MASTICATORIES

Nicotiana tabacum sp. (Tobacco)

VEGETABLES

Abelmoschus esculentus (L.) Moench. (Lady's finger)

Carica papaya L. (Papaya)

Colocasia esculenta L. Schott. (Taro, Cocoyam)

Coccinia grandis L. Voigt. (Ivy gaurd)

Cucumis melo L. (Sweet melon)

Cucumis sativus L. (Cucumber)

Cucurbita moschata Duch (Musk melon)

Lablab purpureus Linn. (Indian butter bean)

Lagenaria siceraria Standl (Bitter bottle gourd)

Musa sp. (a variety of plantain)

FRUIT TREES

Mango, Jackfruit, Banana, Guava, Papaya, Citrus plants.

References

Bose, N. K., 1973, *Some Indian Tribes*, New Delhi.

Bose, S., 1961, 'Land-use Survey in a Juang village', *Man in India* 41(3):172-83.

Brokensha, D., D. Warren and O. Werner (eds.), 1980, *Indigenous knowledge systems and development*, Lanham: University Press of America.

District Statistical Handbook, Kendujhar, 1993.

Elwin, V., 1948, 'Notes on the Juang', *Man in India,* 28(1-2):1-143.

Fernandes, W. (ed.), 1996, *Drafting a People's Forest Bill*, New Delhi: Indian Social Institute.

Gadgil, M., R. Guha, 1992, *This Fissured Land. An Ecological History of India*, Government of Orissa, Delhi: Oxford University Press.

Guha. R., M. Gadgil, 1996, 'What are Forests for?', In: W. Fernandes (ed.), *Drafting a People's Forest Bill*, New Delhi: Indian Social Institute, pp. 33-67.

Kothari, A., 1997, *Understanding Biodiversity*, New Delhi: Orient Longman.

Lévi-Strauss, C., 1966, *The Savage Mind*, Chicago: Chicago University Press.

Mohanty, K. K., 1992, 'The Juang Origin Myth', *Adibasi*, vol. XXXII, no. 4.

Office of the Principal Chief Conservator of Forests, *Decade of Forestry in Orissa 1981-90*, Orissa, Bhubaneswar.

Orissa Gazetteer, Kendujhar, 1986, Department of Revenue, Govt. of Orissa.

Patnaik, N., 1964, 'Caste in Formation among the Juang of Orissa', *Man in India*, 44:22-30.

Patnaik, N., 1986, 'The Juang of Orissa : their work and food intake, demography and fertility', *Adibasi* 26(3):25-65.

Patnaik, N., S. Bose, 1976, *An Integrated Tribal Development Plan for Kendujhar District, Orissa*, Hyderabad: NICD.

Rout, S.P., 1969, 'Handbook on the Juang', *Adibasi,* 11(1-2):1-97.

Rappaport, R.A., 1971, 'Nature, Culture and Ecological Anthropology', In: H.L. Shapiro (ed.), *Man, Culture and Society*, Oxford: Oxford University Press.

Seeland, K., 2000, 'What is Indigenous Knowledge and Why Does it Matter Today?' In: Seeland, K., F. Schmithüsen (eds.), *Man in the Forest — Local, Knowledge and Sustainable Management of Forests and Natural Resources in Tribal Communities in India*, New Delhi: D.K. Printworld, pp. 33-47.

Von Liebenstein, G.W., 1997, Networking for Indigenous Knowledge: Towards a strategy to promote indigenous knowledge in the development enterprise, (CIRAN/NUFFIC) The Netherlands.

White, L., 1959, *Evolution of Culture*, New York: McGraw Hill.

Warren, Michael D., 1997, *The Incorporation of Indigenous Knowledge into Project Implementation for the Development of Indigenous Peoples* (CIKARD), Iowa State University, USA.

4

Biodiversity in Juangpirh, Orissa

B.P. Choudhury

Introduction

THE term biodiversity has become very popular among the scientific establishment although the concept has been around for some time. People of yore, used to stay in places where biological diversities were remarkable because of the availability of various material resources needed for the sustenance of life. On the other hand, their lifestyle was more or less regulated by the environment of that region. Hence, biodiversity is a characteristic regulator of life. At present, biologists, and the general public of various countries have become more conscious about biological diversity. Consequently, the concept of management, judicious exploitation and conservation of biodiversity is critical.

Destruction of biodiversity by anthropogenic factors has accelerated the concept of effective conservation strategies. At present, the global concept of biodiversity embraces a broad spectrum of variations in living forms, both in phenotypic plasticity, as well as genotypic make-up. This, in turn, includes intra- and inter-specific relationships, variability in the structure of niches, and adaptations by organisms to ecological patterns for their survival. Other facets include the diversity of chemicals synthesised by organisms in a specific habitat, and the various metabolic pathways concerned in their life cycles.

In Orissa, places like, 'Bhitar-Kanika Wildlife Sanctuary' (under the newly formed Kendrapara District); 'Similipal Biosphere Reserve' (Mayurbhanj District); 'Satkosia Gorge' (Dhenkanal District); hilly terrains of Kendujhar, Koraput, Ganjam districts; have been identified as rich repositories of biodiversity. Most, other regions are under the destructive spell of anthropogenic pressure. Due attention should be given to check genetic erosion of the biota. In the present context, the lifestyle of the Juang in Gonasika, Kendujhar District will be dealt with. Concurrently, the restoration of biodiversity for a peaceful and harmonious existence of various life forms will be taken into consideration.

Geography and topography

Kendujhar is one of the thirteen districts in the northern part of Orissa. It is situated at the height of 1200 meters above sea level. Topographically, the district is broadly classified into three divisions, namely: (a) the plain regions of the Baitarani river belt, (b) the hilly regions of Bhuinyapirh and Juangpirh, and (c) the sub-mountainous regions of the western and northern parts. The river Baitarani originates from the Gonasika hills. Apart from a few urban areas there are quite a large number of tribal and rural villages. The district has mostly red laterite soil with a few scattered areas of sandy humus soil. On both sides of the river Baitarani there are rocky beds interrupted by sandy rocks. The forest floor usually contains rich clay soil in humus deposit.

Climate

In general, there are four seasons namely:

(i) Hot-dry season,

(ii) Hot-humid season,

(iii) Cool-humid season,

(iv) Cool-dry season.

The monsoon starts in June and extends up to September. The average annual rainfall is 1534.5 mm. It has been observed that there has been a decline in the annual rainfall from the year 1993 onwards and it is often erratic. One of the major reasons for this decline of rainfall is claimed to be the high degree of deforestation due to rapid urbanisation. The maximum temperature recorded is 41°C in May, being the hottest month, while December is the coldest month when temperature comes down to 13°C, but rarely below. The relative humidity of the atmosphere is high throughout the year.

In general, the vegetation can be classified into three categories, namely: (1) Forest, (2) Grassland, and (3) Cropland. The forest vegetation, as identified by Champion and Seth (1968) is mainly of semi-evergreen forest, and moist-deciduous forest.

Material and method

The congenial topography as well as climatic factors are favourable for the growth and development of various species of flowering plants as well as ferns exhibiting a remarkable spectrum of phenotypic plasticity. The hilly terrain along with the rich vegetable wealth provide a favourable niche for the tribal population such as the Juang and the Bhuinya. The lifestyle of the Juang in relation to various plant species will be dealt with in this paper. The biodiversity of the Gonasika region, where the concentration of Juang is high, is remarkable. Haines, the colonial conservator of forests of Bihar and Orissa did not provide adequate information regarding the flora of this region in his treatise (1921-25). Detailed accounts of ethnobotanical aspects are rather lacking from this period. Realising the inadequacy of Haines, Mooney (1950) brought out a 'supplement', based on floristic explorations in the western zone of Orissa with little material on the flora of Kendujhar District.

Hence, an ethnobotanical study was launched under the auspices of the Social Science and Development Research Institute (SSADRI), Bhubaneswar. Through intensive ethnobotanical studies, the biodiversity of the flora in relation to the lifestyle of the Juang was documented. A good number of plant species related to ethnobotanical aspects have been collected, identified and preserved in the form of herbarium specimens following the standard herbarium methodology. The nomenclature of all plant species has been updated in confirmity with the Rules and Articles of the International Code of Botanical Nomenclature (Tokyo Code, 1994).

Analysis of vegetational types

SEMI-EVERGREEN FOREST

The semi-evergreen forest is extant in the hilly tracts and is usually confined to gorges. Its manifestation is unique in Baitarani *munda* near Gonasika village. The major constituent elements of the top canopy are:

Alstonia scholaris, Anthocephalus chinensis, Artocarpus heterophyllus, Bridelia retusa, Diospyros malabarica, D. melanxylon, Ficus religiosa, F. benghalensis, Lagerostroemia parviflora, Litsea glutinosa, Michelia champaca, Mangifera indica, Polyalthia longifolia, Pongamia pinnata, Syzygium cumini, Tamarindus indicus, etc. These are associated with robust climbers like *Combretum roxburghii,Calycopteris floribunda, Bauhinia vahlii, Dioscorea pentaphylla, Jacquemontana paniculata, Ventilago calyculata, Millettia extensa.*

The second storey is composed of bushy shrubs such as *Lantana camara, Woodfordia fruticosa, Melastoma malabaricum, Ixora pavetta, Hyptis suaveolens, Helicterie isora, Flemingia chapper, Allophylus serratus, Dendrocalamus strictus, Clerodendrum viscosum, Cipadessa baccifera, Adhatoda zeylanica,* etc. The common associates are: *Abrus precatorius, Asparagus racemosus, Atylosia scarabaeoides, Cassytha filiformis, Cocculus hirsutus, Hemidesmus indicus, Pergularia daemia,* etc. This close association of bushy shrubs and twinner/climbers provide a congenial ecological niche for many animals.

The ground flora is very rich and embraces diverse types of elements. Of these, *Zornia gibbosa, Vicoa indica, Tridax procumbens, Synedrella nodiflora, Sida acuta, Scoparia dulcis, Rungia pectinata, Plumbago zeylanica, Cyperus distans, Mollugo pentaphylla, Justicia betonica, Fimbristylis acuminata, Desmodium triflorum, Alloteropsis cimicina, Pennisetum pedicellatum, Paspalidium flavidum* deserve special mention.

MOIST DECIDUOUS FORESTS

The major tree species constituting the top canopy are: *Xylia xylocarpa, Terminalia chebula, Semicarpus anacardium, Shorea robusta, Careya arborea, Dillenia pentagyna, Gardenia gummifera, Gmelina arborea, Lannea coromandelica, Ochna obtusata, Madhuca indica.* These are associated with robust climbers similar to those of the semi-evergreen elements. The shrubby elements forming the second strata are *Woodfordia fruticosa, Vitex negundo, Phoenix acaulis, Lantana camara, Glycosmis arborea, Toddalia asiatica, Pennisetum pedicellatum.* The twinners like *Hemidesmus indicus, Coccinia grandis, Merremia tridentata, Shuteria densiflora, Melothria heterophylla* are found in abundance.

The floristic composition of the ground flora is almost similar to that of the semi-evergreen forest. However, most of it is *heliophilus*. Notable taxa are *Croton bonplandianus, Sida cordata, Alternanthera pungens, Amaranthus spinosus, Argemone mexicana, Vernonia cinerea, Blumea lacera, Cyperus iria, Cassia tora, Cleome viscosa, Eragrostis coarctata, Sporobolus indicus, Perotis indica, Chloris barbata, Fimbristylis acuminata, Cyperus rotundus, Abutilon indicum.*

In the open marshy places one usually finds *Ammannia baccifera, Cyperus iria, Phyla nodiflora, Coldenia procumbens, Cyperus diformis, C. nivexs, Fimbristylis aestivalis*. The herbaceous species come under two categories, namely (a) monsoon flora, and (b) non-monsoon flora. The monsoon flora appears just after a shower of rains at the advent of the rainy season. Species like *Desmodium triflorum, Croton bonplandianus, Kyllinga triceps, Cyperus distans, C. nutans, Melochia corchorifolia, coldenia procumbens, Corchorus aestuans* appear at the beginning of the rainy season, bloom, set seeds and then disappear at the end of winter. The second category comprises of taxa like *Jatropha gossypifolia, Maytenus emarginatus, Lepidagathis hamiltoniana, Elephantopus scaber, Flacourtia indica*.

The floristic elements of the hill slopes are often destroyed by the Juang due to shifting cultivation. The foodgrains procured by this process are not sufficient for the whole year. In their anthropo-geographical work on micro-level planning in Kendujhar district, Patnaik and Bose (1976) have duly indicated how eager the Juang are in giving up shifting cultivation and adopting paddy cultivation in the valley-bottom.

In the semi-evergreen forest quite a number of epiphytes add to the biodiversity in this region. Plants like *Bulbophyllum triste, Cymbidium aloifolium, Dendrobium transparens, Dendrophthoe falcata, Habenaria commelinifolia, Oberonia ensiformis* deserve special mention. However, quite a number of orchids (mostly in the epiphytic category) such as *Liparis vestita ssp. seidenfadonii, Malaxis densiflora, Habenaria Panigrahiana, H. diphylla, H. stenopetala, H. plantaginea* which have been reported by Misra (1980, 1981, 1982, 1986 and 1987) could not be traced during the present investigation. These plant species are very sensitive and any change in their niche and microclimate affects their life cycle.

Consequently, if such changes continue, these orchids may disappear. Apparently the Juang do not interfere with this plant community. However, due to large-scale deforestation they are disappearing. The Juang, after consuming their stock of rice, use rhizomes, bulbs, some fruits, and bulbils to meet their food demands.

Relevant plants in Juang life

The Juang tribe was studied for the first time by Samuells (1856). It is known from the prevailing myth that the origin of the Juang is traced from the mythical ancestors *Rusi* and *Rusiani* since the Vedic age. The myth has been dealt with by Mohapatra (1989) and Mohanty (1992), as well as Elwin (1952) regarding the origin of this tribal group. Detailed information has been provided by Patnaik (1986) on the daily and annual work rhythm, the food habits, and common medico-religious beliefs of the Juang.

The Juang use the rhizomes of *Colocasia esculenta*; fruits of *Cucumis sativus*; the roots of *Impomoea batatus*; fruits of *Lagenaria siceraria;* rhizomes of *Dioscorea belophylla;* bulbs of *Dioscorea bulbifera.* Oil is extracted from the seeds of *Ricinus communis, Schleichera oleosa, Pongamia pinnata, Guizotia abyssiniea. Brassica campestris* is used by the Juang for various purposes. *Microtyloma uniflorum, Vigna mungo, Vigna unguiculata* are also used. Besides these food plants, they depend upon *Bauhinia vahlii, Butea parviflora, Careya arborea, Diospyros malabarica* for fibre. Similarly, roots of plants like *Achyranthus aspera, Alangum salvifolium, Amaranthus spinosus, Boerhavia diffusa, Cissampelos pareera, Elephantopus scabre, Toddalia asiatica, Sida acuta* and *Sida cordifolia* are extensively used for various purposes by the Juang. Similarly, seeds and fruits of *Terminalia chebula, T. bellirica, Strychnos nux vomica, Phyllanthus emblica, Shorea robusta, Mucuna pruriens, Smilax zeylanica* are used for various purposes. Roots of *Rauwolfia serpentina, Plumbago zeylanica, Calotropis procera* are mostly used to cure various diseases. Timber and house-building materials are procured from the plants like *Adina cordifolia, Albizia lebbeck, Anogeissus latifolia, Bridelia retusa, Dendrocalamus strictus, Shorea robusta, Xylia xylocarpa, Syzygium cumini, Lagerstroemia parviflora*. The various parts of the plants used in magico-religious rituals are: *Achyranthus aspera, Amaranthus spinosus, Shorea robusta, Dendropthoe falcata, Smilax zeylanica, Plumeria rubra*. The woody parts of *Bridelia retusa,*

Tamarindus indica, Bridelics retusa are used as fuel. The bark of *Careya arborea, Firmiana colorata, Litsea glutinosa, Kydia calycina* is used to make ropes.

The Juang enjoy plentiful resources for the sustenance of their lives and the biological richness of their habitat is impressive. However, adequate strategies should be adopted to protect this rich biodiversity, as their way of life is increasingly endangered by the unsustainable use of India's natural resources.

References

Bose, N. K., 1928, 'Marriage and Kinship among the Juang', *Man in India*, 8 (4):233-42.

Champion, H.G., and S.K. Seth, 1968, *A revised Forests types of India*, New Delhi.

Elwin, V., 1952, *Tribal myths of Orissa*, Oxford: Oxford University, Press.

Haines, H.H., 1921-25, *The Botany of Bihar and Orissa*, London.

Misra, S., 1980, 'Addition to the orchidaceous flora of Orissa', *Bull. Bot. Surv. Ind.*, 22, (1-4) :147-56.

Misra, S., 1981, 'Two new orchids from Orissa', *Blumea,* 27: 213-16.

Misra, S., 1982, 'Addition to the Orchidaceous flora of Orissa-II', *J.Orissa. Bot. Soc.*, 4 (1) : 23-25.

Misra, S., 1986, 'Two new orchids from Orissa; India', *Nord. J. Bot.* 6 (1): 25-29.

Misra,S., 1987, 'Addition to the Orchidaceous flora of Orissa-III', *J. Orissa. Bot. Soc.*, 9.9(1): 32-37.

Mohanty, K.K., 1992, 'The Juang origin myth', *Adibasi,* 32 (4):1-8.

Mohapatra, P.K., 1989, 'Myths of Juang — An anthropological analysis', *Adibasi,* 29 (3-4):21-23.

Mooney, H.F., 1950, *The supplement to the floristic studies of India*, Catholic Press, Ranchi.

Patnaik, N., 1986, 'The Juang of Orissa, their work and food intake demography and fertility', *Adibasi,* 26(3): 25-65.

Rout, S.P., 1969, 'Hand book of Juang', *Adibasi,* 11 (1-2):1-97.

Samuells, E.A., 1856, 'Notes on a forest race called Puttooas or Juanga inhabiting certain of the tributary mehals of Cuttack', *Journal of the Asiatic Society of Bengal*, 25.

Senapati, N., D.C. Kumar, 1977, *Orissa District Gazetteers: Kendujhar*, Orissa Govt. Press, Cuttack.

Table: Oriyan English and Kuvi names of the botanical plants found in the area

Botanical Name	Oriya Name	English Name	Kuvi Name
Abrus precatorius L.	Kaincha	Crab's *eye*	Runja
Acacia lenticularia	Kanta sirisa	-	-
Acalypha indica L.	Indra marisa	Indian acalypha	Naktichana
Adina cordifolia (Roxb.) Hook.f. ex. Bran.	Kuruma	-	Halanda mara
Aegle marmelos (L.) Corr.	Bela	Bael tree	Bela mara
Ageratum conyzoides L.	Pokasunga	Goat weed	Urla
Albizia procera Benth.	Dhala sirisa	White siris	Sirisan
Alocacia indica Schott.	Manasaru	Giant taro	-
Aloe vera (L.) Burm.	Ghee kuanri	Indian aloe	Ghee kuanri
Alstonia scholaris (L.) R. Br.	Chhatiana	Dita bark	Sasina
Amorphophallus campanulatus (Roxb.) Bl. ex Dec.	Olua	-	Ola
Anacardium occidentale	Kaju	Cashew nut	Bhalia
Annona squamosa L.	Newa	Custard apple	Sitaphala
Anogeissus latifolia (DC.) Bedd.	Dhaura	Axle-wood tree	Dhoa mara
Argemone mexicana L.	Odasamari	Prickly poppy	Agara
Artocarpus lakoocha auct Roxb.	Jeuta	Monkey jack	Jeuta mara
Azadirachta indica A. Juss.	Nimba	Neem tree	Nima mara
Barringtonia acutangula (L.) Gaertn.	Hingala		Hinjala mara
Bauhinia retusa	Choari	-	-
Bauhinia Vahlii Wt.& Arn.	Sialilata	-	Paeri
Blumea lacera (Burm.) DC.	Pokasunga	-	-
Boerrhavia diffusa L.	Puruni	Horse-purscane	-
Bombax malabaricum DC.	Simili	Silk cotton tree	Leka mara
Brassica juncea (L.) Czer.	Sorisa	Mustard	Sarsakucha

Bridelia retusa Spreng.	Kasi	-	Kasi mara
Buchanania latifolia Roxb.	Chara	Cuddapah almond	Chara mara
Bursera serrata Coleb.	Nimburu	-	Nimbrumahi
Butea monosperma (Lam.) Taub.	Palasa	Flame of the forest	Palasa
Butea superba Roxb.	Palasa lata	-	Nai palasa
Cajanus cajan (L.) Millsp.	Harada	Pigeon pea	Kaanga
Callicarpa indica	Boda	-	-
Calotropis gigantea (L) R.Br.	Arakha	Giant-milk weed	Arakmara
Calotropis procera R.Br.	Arakha	-	Arakmara
Carica papaya L.	Amrutabhanda	Papaya	Ambrut
Caryota urens L.	Salapa	Sago palm	Mada mara
Cassia fistula	Sunari	Indian Laburnum	Bundeni mara
Cedrela toona Roxb.	Tun	-	Tun mara
Cipadessa fruticosa	Nalabali	-	Nahalabeli
Cissampelos pareira (DC) Forman	Akanabindi	False Pareira Brava	-
Cissus quadrangularis L.	Hadabhanga	The edible stemmed vine	Pdenulaha
Citrus aurantifolia (Chris.) Swin.	Lembu	Lime	Lembu
Citrus aurantinum L.	Khata kamala	Biggord orange	Kambla
Citrus limon (L.) Burm.f.	Jambhila	Rough Lemon	Narji
Citrus maxima (Burm.) Merr.	Batapi	Shaddock, Pummelo	Batei
Citrus reticulata Blanco.	Santara	Loose-skinned orange	Karuna
Clerodendron infortunatum auct non. L.	Genguti	-	Kharkhari
Clitoria ternatea L.	Aparajita	-	Aparajita
Coffea benghalensis	Banamali	-	-
Combretum decandrum	Atundi	-	-
Cucurbita maxima Duch.ex Lam.	Kakharu	Pumpkin	Kumda

Botanical Name	Oriya Name	English Name	Kuvi Name
Curcuma aromatica Salisb.	Bana haladi	Wild turmeric/Yellow zedoary	Anahinga
Curcuma longa L.	Haladi	Turmeric	Hinga
Curcuma zedoaria Rosc.	Palua	Arrowroot	Batra
Cycas circinalis	Odasamari	-	-
Cynodon dactylon Pers.	Doobghas	Bermuda grass	Doob
Cyperus rotundus L.	Mutha	-	-
Dalbergia lanceolaria L.f.	Chakunda	-	Sajanapati
Datura stramonium L.	Dudura	Thorn apple	Dudura
Dillenia indica L.	Waou	Elephant apple	Waou mara
Dillenia pentagyna Roxb.	Raee	-	Rael mara
Diospyros melanoxylon Roxb.	Kendu	Ebony	Kendu mara
Eleusine coracana (L) Gaertn	Mandia	Finger millet	Mande
Embelia robusta Roxb.	Nunnunia	-	Baibidanga
Erythrina suberosa Roxb.	Paladhua	-	-
Eugenia jambolana	Jamu	Black berry	Jamumara
Euphorbia antiquorum L.	Siju	-	Hipalu
Ficus benghalensis L.	Bara	Banyan tree	Bada
Ficus glomerata Roxb.	Dimiri	-	Kharsan
Ficus religiosa L.	Aswattha	Pipal tree	Bada/Jari
Fleurea interrupta (L.) Gaud	Bichhuati	-	Hengulimara
Gleistanthus collinus	Karada	-	-
Gmelina arborea Roxb.	Gambhari	Coomb tree	Gambar mara
Gossipium hirustum L.	Kapa	Cotton	Parti
Grangea maderaspatana (L.) Poir. in Lam.	Agnikumari	-	-
Grewia tiliaefolia Vahl.	Dhaman	-	Daman mara
Guizotia abyssinica	Alasi	-	Kuyan
Hibiscus rosa sinensis L.	Mandar	China rose	Mandar

Holarrhena antidysenterica (Roth.) wall.ex Dc	Kuduchi	-	Palche mara
Imperata arundinacea	Sabai	-	Wika
Indigofera cassioides Roth ex Dc.= *Indigofera pulchella* Roxb.	Nila	-	Giliri Kucha
Ipomoea batatas (L) Lam.	Kandamula	Sweet potato	Ranikuna
Jatropha curcas L.	Amarijada	-	-
Kydia calycina	Banakapasia	-	-
Lagenaria piceraria (Mol.) Standl.	Lau	Bottle gourd	Pita
Lantana camara L.	Nagabairi	Lantana	Pungalaha
Lxora parviflora Vahl.	Katharangani	Jungle flame ixora	Rangni
Madhuca longifolia (Koenig)	Mahua	Mowra butter tree	Irpimara
Mallotus philippinensis (Lam) Muell.	Kamalagundi/ Sindura	-	Sendur mara
Mangifera indica L.	Amba	Mango	Amba
Michelia champaca L.	Champa	-	Japimara
Murraya koenigii (L.) Spreng.	Bhusuranga	Curry leaf tree	Bhersinga
Musa paradisiaca L.	Kadali	Banana	Tadi
Ocimum basilicum L.	Sugandha	Sweet basil	Landa babuli
Oldenlandia corymbosa L.	Gharpodia	-	-
Oryza sativa L.	Dhana	Paddy plants	Kulinga
Ougeinia dalbergioides	Bandhana	-	Bandan mara
Pandanus fasicularis Lam.	Kia	Screw pine	Kia
Panicum miliare Lam.	Suan	Little millet	Kosala
Phoenix acaulis	Bhuin khajuri	-	Mila Sindhi
Phyllanthus emblica syn:*Emblica officinalis* Gaertn.	Amla	Indian Goose Berry	Surga mara
Piper nigrum L.	Golamaricha	Black pepper	Golamaricha
Polinidium angustifolium	Bagali	-	-
Psidium guajava L.	Pijuli	Guava	Jambu
Pterocarpus marsupium Roxb.	Piasala	Gum kino tree	Bija mara

Botanical Name	Oriya Name	English Name	Kuvi Name
Ricinus communis L.	Jada	Castor oil plant	Kabang
Rosa indica Linn.	Golap	Rose	Golap
Santalum album L.	Chandan	Sandalwood tree	Chandan
Sapindus emarginatus Vahl.	Ritha phala	-	Sikdikai
Schleichera oleosa (Lour) Oken.	Kusum	-	-
Semecarpus anacardium L.f.	Bhalia	Marking nut tree	Banabhalia
Shorea robusta Gaertn.f.	Sala	Sal tree	Sargi mara
Smilax macrophylla Roxb.	Kumbhatua	-	Rajdanti
Solanum surattense Burm.f.	Ankaranti	-	Bheji
Sorghum vulgare	Bajra	Bajra	Kangu
Strychros nux vomica L.	Kochila	-	Pitana
Syzygium cumini (L) Skeels	Jamu	Black berry	Jamu mara
Tamarindus indica L.	Tentuli	Tamarind	Leli mara
Tectona grandis L.f.	Saguan	Teak	Saguan
Terminalia bellirica (Gaertn) Roxb.	Bahada	-	Baena
Terminalia tomentosa (DC) W&A.	Asana	Asan/Sain	Sahaj mara
Themda arundinacea	Chhana	Thatching grass	Ladan
Thysanolaena maxima O. Ktze. *Syn: Thysanolaena agrostis* Nees.	Phooljhadu	Broom grass	Satra
Tinospora cordifolia (Willd) Miers.	Guluchi	-	Guluchi
Trema orientalis	Kharkas	-	-
Tridax procumbens L.	Bisalyakarani	-	Bisalyakarani
Vanda roxburghii	Rasna	-	Madanga
Vigna indica L.	Jhudunga	Cow pea	Jhudunga
Vitex nigundo L.	Begunia	Indian privet	Nirgundi
Wendlandia tinctoria	Tilai	-	-
Woodfordia fruticosa Kurz.	Dahiphula	-	Dhatuki
Wrightia tomentosa Roem.&Sch.	Khirua	-	-
Xylia xylocarpa (Roxb.) Taub.	-	-	Dengen/Kangda

5

Food, Folk and Forest

Alok Rath

Introduction

THERE is little doubt that tribal groups expend more energy on food procurement than anything else. This is especially because of the vulnerability of their sources of food. Unlike sex, hunger is a recurring factor and cannot be tolerated beyond a limit. Yet there are social factors which regulate the food-getting and intake of an individual in different societies. In a Hindu society, a number of religious occasions exist where fasting is strictly observed and, in some cases, even water is forbidden. The point is, that even though nutrition is a fundamental biological process, it is regulated by social principles. An orthodox Hindu does not eat beef, a Brahmin non-vegetarian food, a Muslim does not eat pork. Such restrictions influence specific behavioural patterns which become group identification marks. The Juang are one of the most primitive tribal communities in Orissa, spending almost all their energy throughout the year in food procurement. Being primarily swiddeners, their annual activities revolve around the process of shifting cultivation and other subsistence activities.

Concept of food

Food is vital for survival and it comes mostly from the forest. Swiddens are carved out from the forest: wild fruits, roots, greens, and game are all gifts of the forest without which Juang survival would not be possible.

Traditional modes of food procurement

HUNTING

For generations, the Juang have been dependent on their surrounding forests for their food requirements. Not long ago, their subsistence revolved around hunting game and the collection of wild plants. Nineteenth-century ethnographic investigations by Dalton (1872) and Hunter (1877) suggest that the Juang subsisted primarily on wild game, which is no longer the case today. Hunting as a co-operative effort still has a social meaning. Meat used to be a welcome change from the rigours of the arduous swidden operations. The communal hunting (*paridhi*) expeditions still take place during March-April after observing the first-fruit-rite for mango (*amba nua*) and continue for about a month. It is also the period during which the plants are slashed and burnt in the swiddens. After that, the ashes are mixed with the soil to increase fertility; so it is the ideal period for concentrating energy on hunting. An elaborate ritual process is attached to communal hunting (for details see Rout 1969:37-39). Even though this period is attributed to communal hunting, individual and small group efforts are also common. In fact, the game and the weapons used to hunt it determine the efforts in preparation. Hunting of big game, such as wild boar and bear, requires a big group effort, while for hunting and/or trapping small game such as rabbit, or jungle fowl, individual efforts may be sufficient. When the first rains come and the demanding agricultural operations require a massive labour input, group hunting stops, although individual hunting may continue.

Communal hunting is the medium through which the bravery of individuals is socially recognised. Only daring and courageous individuals (usually one member from each family) participate. Mere participation in a hunting trip makes one recognised as courageous in the village. But the highest esteem for bravery is reserved for the individual who actually kills an animal. He is referred to as the *raja* or king. It brings him instant social prestige and reverence and he is considered auspicious. Such fame results in positive rivalry to excel in hunting. Sharing of the meat is what brings the Juang together in a hunting trip. Except for the actual hunter who gets an additional share, all other participants get equal shares, irrespective of the extent of their involvement in the chase. Besides the individual shares, there are family shares distributed equally among all the families in the village. In addition, a specified share is also kept aside for the ritual elders. Kinship does not play any role in such an effort, but the communal bond in the village is reinforced. Group hunting generally excludes older Juang who are no longer considered enough, but includes capable 'recruits' who get a chance to hone their hunting skills. The additional share that the actual hunter gets goes to his maternal uncle who, in turn, gives him an arrow and blesses him for more success in hunting. The auspicious hand of the hunter is recognised and his economic importance to the community is thus valued. His own original share of meat is further divided into the families of his own lineage (Kutumali), and that of other Kutumalis in the village. Since communal hunting does not always meet with success, the actual hunter becomes a hero among the fellow villagers. In many cases, the big game is not killed on the spot after being hit and may die in the territory of another village. In such a case, the game has to be shared between the two villages, with the bigger share going to the hunting side. Such customary rules not only regulate inter-village relations, but provide a distributive mechanism of meat. This applies equally to small group hunting and individual hunting. In the case of small group hunting, which I would call 'pair-hunting' because of the number of hunters involved, new relationships emerge in the community. In sports, we find phrases such as 'forward pair' in soccer, 'opening combination' in cricket, because of the higher percentage of success where a strong bond exists between two individuals. Similarly, a higher percentage of hunting success develops over a period of time if two Juang develop a relationship as strong as kinship ties. Their relationship is, based on mutual hunting auspiciousness. In the case of individual hunting, meat is also redistributed equally among all the families in the village if the hunter has killed big game.

AGRICULTURE

The mainstay of Juang subsistence is their production from swidden cultivation. The entire year is concentrated in swidden operations. The entire workforce of a family is engaged in earning food from the hill-slopes. Inter-village and intra-village kinship, as well as other relationships revolve round this subsistence activity.

The very nature of a dyadic division of Juang society into *kutumb* (agnatic) and *bandhu* (loosely affinal) villages is believed to be the result of the pursuance of swidden cultivation (Dash 1988:22). This arduous form of agriculture requires a large-scale labour input that is difficult to meet by a single family, hence the search for labour from the agnatic circle. This led to the formation of uniclan, exogamous villages. Even today, when many Juang villages are inhabited by members of more than one clan, we can find co-operative labour exchange among agnates in agricultural operations. The agnatic bond is further strengthened through the moral obligation of providing grains — by a surplus production village to a deficit production village — on nominal interest in times of need. A married daughter may help her father harvesting crops if she has finished her own harvesting. Usually, a group of girls from a *bandhu* village visit another one and participate in harvesting there. During their visit, they perform *changu* dances during the nights throughout their stay. This symbolises the meeting and mingling of *bandhu* boys and girls, in many cases resulting in selecting mates. Such occasions give a chance to prospective partners to get to know each other. The visiting team is paid in grains for its labour in the field.

At the start of a new clearing, the villagers gather at their community centre (*majang*) to seek divine consent and blessings for a successful cultivation year. Done through their village priest, Boita or Dehuri, this act reassures the Juang of a bountiful harvest. The distribution of swidden plots which are communally owned is periodically done by the village priest whose decision everyone accepts, for he speaks with a divine voice. This process maintains discipline in the village. The priest takes note of someone getting an inferior plot in one year. The same individual is given a better plot during the following cycle. Once a plot is allotted to one Juang he becomes the owner of it for a swidden rotation period (usually for three years) even if he does not cultivate the plot for the entire rotation cycle. He may, however, permit someone else, usually an agnate, to cultivate the plot on his behalf, if he himself is unable to do so.

Swidden cultivation, determines the division of labour among the Juang to a large extent. Felling of virgin forest for fresh cultivation requires hard labour, and hence only men are engaged in such activity. Once the hard work is done, the women remove the remaining thinner stumps from the plots. The next job of ploughing is again done by men. The Juang myth says that Juang women are not allowed by their supreme ancestors *Rusi* and *Rusian* to touch the handhoe, lest the soil loose its fertility. The women thus carry out the next act of degrassing and weeding, after which the men protect the growing and standing crops from animals. The final phase of harvesting is again done mostly by women. If we look at the calendrical work engagement in the swidden, it is interesting to note that all the stages in swidden operations are done alternately by men and women. In all such activities, young Juang (10-14 years) are engaged as trainees to subsequently take charge of their own responsibility when they establish separate households after marriage.

This division of labour in the swiddens also influences other activities for both men and women. During felling and burning when men are busy in the swiddens, the women engage in forest activities that include the collection of minor forest produce and fuelwood. When the women are removing small stumps and weeding, the men usually go to fetch bigger fuelwood (logs) and deep-rooted tubers such as *Jalajang* (*Dioscorea* sp.). The women, however, have to do the cooking and provide food for the men working in the swiddens.

SOCIO-CULTURAL LIFESTYLE

A number of social and cultural activities go on side-by-side with this division of labour. Fuelwood collection in the forest gives the women the opportunity to discuss various family and community matters. For young, unmarried girls both forest activities and working in the swiddens include creative in singing and song composition. Most of the Juang songs are either nature-related and/or about love and romance composed mostly by the Juang youth. When someone is composing and singing the opening stanza, someone else would add another stanza and this goes on for a while. This is the reason why many Juang songs are lengthy and repetitive. The forest and its surroundings inspire them to compose songs coming straight from their heart.

Juang men, especially the young, go to sleep in the swidden fields at night to watch and protect the standing crops from being destroyed by wild animals. Usually, the owners of two or three adjacent plots sleep together in a common field-hut, leading to close friendship. Apprenticeship of a prospective shaman (*raulia*) also commences during this period. The teacher shaman asks his pupil to sleep with him in the field-hut where the latter's training takes place on the pretext of crop watching. This way, the secrecy of *raulia*-apprenticeship is maintained until the training is over, and the trainee is declared a shaman by his tutor.

The most important form of Juang cultural activity, the *changu* dance, takes place during and following harvesting. Occasionally, *changu*-dancing is performed by girls and boys of the same village at any time of the year. This is the most relaxing period for the Juang and they thoroughly enjoy it by drinking and dancing throughout the nights. As has already been mentioned, girl teams from *bandhu* villages visit other *bandhu* villages during harvesting. After harvesting, boy teams pay visits to the girls' *bandhu* villages. Gifts, mainly food, are exchanged and relations are established leading to marriage in various forms such as arranged marriage, marriage by capture, or elopement. Juang marriage is directly linked with their swidden food production. In the bride-price system, the boy has to give 60 kg (3 *khandi*) of dry rice and 40 kg (2 *khandi*) of paddy to the girl's family, and also has to spend a lot in feeding his own villagers. Marriage takes place only after the harvest. If the yield is less and insufficient for consumption at home, the marriage is postponed until the following year. Most of the private life of a young Juang couple takes place during this period. It is a convenient period for them to be engaged in amorous moments in the forests. This also helps them to stay away from the young children who are still dormitory members in the village.

A Juang child gains knowledge about his own society and environment in the swidden and forest, as most of his infancy and early childhood is spent there in his mother's couch (made with her saree). The child is caressed and lullabies are sung which are, at times, spontaneonsly composed and refer to Nature and deities. When the child grows older, it receives education relating to Juang myths, legends and folktales, which give him a clear understanding of his society and environment. Most Juang proverbs and riddles are connected to their swidden cultivation practices. In fact, a large part of their anthology is connected to their means and modes of food procurement and the child comes to know them all. The entire gamut of indigenous knowledge relating to survival techniques is learnt by the child through mock acts, after observing the activities of the elders. When Juang children play in the villages, their games include hunting, hedge-making, ploughing, *changu*-dance and are accompanied by sweet and rhythmic songs. By the time they become old enough to be economic contributors to the family household, a high degree of skill in their subsistence activities (and other socio-cultural activities) is already attained.

Collection of forest produce

Many aspects relating to forest activities have already been highlighted whilst discussing swidden cultivation. A strong sense of village territoriality is notable when it comes to forest resources. The collection of timber, fuelwood and wild food is strictly restricted to one's own village forest (demarcated by a boundary). An individual will not exploit the forest resources of another village, without seeking and getting permission to do so. Every village respects the boundary of every other village. At times, when a village is short of swidden area, it may seek permission of another village to clear the latter's forest for swidden cultivation for a limited period. In such cases, the favour-seeking village sends one elderly delegate to the land-owning village, usually with gifts of food and liquor. After harvesting, a specific share is given to the owner village and the share is spent in communal feasting.

In the case of a breach of discipline, the traditional village council of the offended village (or individual) reprimands and/or fines the offending village (or individual). But such cases rarely occur. Discipline is maintained in the exploitation of one's own forest resources. Over-collection of any forest item is hardly seen among the Juang. This behaviour is connected to their strong attachment with the forest. Since he/she visits the forest daily, he/she gets fuelwood, wild food, but only as much as is required for a day or two. Some trees that are planted are individually owned by families, but others are communally or collectively owned. In case of communal ownership, there are periodic distributions of fruits, either cooked or uncooked, to all families (McDougal 1963 a:310).

Conclusions

The above short account provides a sketch of Juang dietary needs, habits and behaviour in relation to their society and environment. It also reinforces the point that the Juang are not only forest-dwellers but dependent on it for their survival. Swidden cultivation is their chief source of subsistence and the broad spectrum of human relations are woven around it. Inter-personal, group, intra-village and inter-village relations are directly or indirectly based on subsistence culture. It is a system that has lasted generations. The recent decades have, however, seen changes in this system, some prominent and some hidden, due to the onslaught of modernity.

Modernity in the form of forest administration has hit them the hardest. Reserved forests and Protected forests have been earmarked, thereby affecting the free movement of the Juang in collecting forest resources. Strict laws have shrunk the area for swidden cultivation. The Protection of Wild Life Act prohibits hunting in the forests. These measures have, no doubt, been directed at the protection and better management of forest resources by the state; but has this aim really been achieved? The bottom line is that it has not, largely due to the State's failure in checking illegal logging and poaching by outsiders. Singh (1988:6) notes that after Independence the consumerist extravaganza of the country's urban elite caused the logging of a massive number of trees. Further, the growing industrial demand dating from British times has also been responsible for the rapid thinning of the woods. The result is large-scale deforestation. There is a popular saying, 'trees have started dwindling from the day the Forest Department was created'. In addition to being ineffective in checking deforestation, these measures and other elements of modernity have had serious repercussions on the Juang culture.

Communal hunting was a food-getting mode that not only collected food but maintained village fraternity among the Juang. Now, there is no such ritual aspect to communal hunting. In fact, there is no more communal hunting at all. Because the Juang are afraid of the Forest Department, hunting is clandestinely carried out occasionally by a group of young boys. The game is cooked and consumed in the forest itself. The entire ritual process attached to hunting is gone, thereby weakening food discipline (i.e. the distribution of the meat mentioned earlier). A greater degree of individualism has crept into their lifestyle. The communal mode of hunting and the code of conduct attached to all forms of hunting was the medium through which village resources were controlled and managed.

Restrictions in swidden cultivation, in addition, have forced the Juang to search for food from other sources such as valley-bottom cultivation, upland cultivation, backyard cultivation, wage earning. These are individual-oriented and lack all the group activity that is associated with the traditional modes of production. This has further eroded group relations and has weakened the continuity of the Juang as a community. McDougal (1963a:309) noted that:

> the Juang derive considerable satisfaction from communal work. It is a social event. The persons engaged in work indulge

> in joking, singing and general merriment. At the end of the day, there is an ample feast, anticipated with pleasure, especially during the months preceding the harvest when the food supplies of most families are low.

From McDougal's work and my own empirical experience, it is clear that in the case of swidden cultivation there are communal plots, communal labour and communal production, a communal feast and a common village fund for mutual help. A strong cohesive feature indeed, but fast fading away under today's conditions.

The most prominent example is the lack of adherence to customary laws and traditional leadership. In the past, the post of the village religious head (*Boita*) was a matter of prestige and a position of respect for the incumbent. He is usually one of the oldest members of the community. Now, every person designated seems to be shying away from it for this post because the post is now thought to be without prestige involving a lot of difficult and can unrewarding responsibilities. The Gonasika village had an interim young *Boita* for 1996 and 1997, while the neighbouring Guptaganga village has not had a priest in 1998.

The traditional village council along with the village elders used to take decisions regarding the swidden activities such as its timing; the direction and size of the forest to be cleared; particular slopes to be avoided; and particular size of trees to be spared. Today, the situation is fast becoming a free-for-all affair, an open access economy. The old feeling that both the forest and the swidden are complementary, has worsened (due mainly to large-scale deforestation) partly because the young do not listen and do not follow the rules of Nature. The youth have their answer, too. They say, since the non-tribal neighbours (Gauda, Pano) are cutting down trees and clearing large patches for swidden cultivation, what is the meaning of following their traditional way? Traditional leadership is no longer the all-important medium for justice dispensation, because of the power of the state and its laws are easily manipulated by the rich and the influential.

The simple and hard-working Juang is dubbed by mainstream society as 'lethargic'. Lazy the Juang has indeed become, because of the lack of communal cohesion and group activity that once energised him. It is sad to know that *changu* dancing, the only form and avenue of Juang entertainment, is no longer a regular event, even during harvesting. There are no frequent visits of *bandhu* boys and girls. The lack of such visits has killed the enthusiasm in the *changu* drum-beating competition. Today, a majority of Juang boys do not know how to beat *changu,* and the girls do not know how to dance. Consequently, the soul-raising Juang songs are also unknown to the Juang youth. They are only on the lips of old Juang. An institution is thus threatened and as it was also a medium of food exchange, this social obligation would lose its significance if the practice dies out.

The traditional modes of food-getting were spheres in which the Juang child developed its understanding about his/her society, culture and environment and acquired knowledge regarding indigenous modes of natural resource use and management, imparted through rituals, taboos, oral literature, and above all, a practical involvement in the food-getting operations. Today, hardly any Juang child knows much about this tradition. The situation is the result of a lack of involvement and interaction between old people and youth. Lamenting poetically through a modern song, the old Juang pour their heart out at the industrial civilisation for being responsible for disintegrating their society, and taking them to the brink of the end.

Juang combs, made out of bamboo and wood with beautiful animal and floral carvings on them, were widely described by Elwin to be a precious cultural feature. The comb was also used as a gift between *bandhu* boys and girls. They were used for *changu*-dancing by fixing the combs on their heads. These combs are no more. In about twenty villages that I have covered during my fieldwork, the comb-making art is still known to only four old Juang in Guptaganga village.

It is interesting to mention here that among the Bhuinyas in Kendujhar, living adjacent to the Juang, *changu* dance has become a thing of the past. Bhuinyas as old as 50 to 60 years do not know their once prized dance anymore. They can only recall having witnessed it in their childhood. Could this be related to swidden cultivation and a lack of communal or group interaction? I believe so. As of today, a majority of Bhuinyas here pursue subsistence activities other than swidden cultivation. The beautiful and gigantic dormitory hall that once acted as the centre of all Bhuinya activities, is now replaced by a government-built, tiny and dingy community hall, where no group activity takes place as the youth organisation is defunct. It is interesting to note here what McDougal had to say about the Juang dormitory:

> The local descent group (the largest village corporate unit) is segmented into small agnatic descent groups, but these exhibit a low degree of solidarity. . . . Because relations of agnatic kinship are not a strong focus for solidarity over time, there must be some alternative, integrative mechanism at the village level, if the local community is to be a solidary unit characterised by stability and continuity. This is provided by the men's house, symbolising village unity, and the grouping of persons associated with it. (1936b : 189-90)

The trend may not be far-off with the Juang. Although the dormitory hall and youth organisation still exist, they have started showing signs of disintegration. There could be many such signs which are hidden or which perhaps have evaded my notice, but are nonetheless in the offing. We know in theory that any change in an aspect of a system brings about a chain reaction in other spheres of that system. A change in the traditional mode of food procurement, therefore, is threatening to weaken and alter the entire cultural fabric of the Juang. As Radcliffe-Brown said,

> A social system . . . has a functional unity . . . which may be defined as a condition in which all parts of the social system work together with a sufficient degree of harmony or internal consistency, i.e. without producing persistent conflicts which can neither be resolved nor regulated. (1952 : 181)

Making an in-depth study among the indigenous populations of the Eastern Ghats in south Orissa, Mohanty finds that a change in the traditional subsistence due to forced migration has eroded their established value system, causing disquiet in their overall living conditions (Mohanty 1991:50). Kapp (1986) discusses at length the social costs of displacement of traditional communities because of industrial activities in the hinterlands. Depriving a community of the life of its choice and convenience, is surely a displacement to disorder.

Besides affecting human relations, a change of subsistence from shifting cultivation to other forms of agriculture, (i.e. valley-bottom paddy cultivation) has had an adverse impact on their food base. It is much too well-known that the mixed inter-cropping in swidden cultivation maintains a rich gene pool! of cereals, millets and pulses, that are ecologically conditioned to grow in adverse conditions (erratic rainfall). The practice of other forms of agriculture leads to the adoption of a paddy monoculture of high-yielding-varieties (HYV) that are susceptible to inadequate or differential rain. Moreover, the adoption of these forms of agriculture has posed a threat to the rich gene pool of traditional crops. For instance, the Juang had about 45 varieties of rice a couple of decades ago now they are reduced to less than a dozen varieties. Joseph (1994) of the Swaminathan Foundation argues that mountain agriculture is ecologically controlled by means of being well adapted to the environment by the presence of different microbial agencies. The phytochemical odour of one plant repels the insects or pests affecting another plant. Genetic heterogeneity lessens prospects of pest attack. This is probably why the Juang do not have any field pest control methods, although they do use indigenous techniques in storing the grains. Last year, I noticed, the standing mustard crop being damaged by insects, something which the Juang felt was quite new to them. The mustard fields are close to the valley-bottom paddy cultivation in which HYVs supplied by the Juang Development Agency are grown. The onslaught of HYVs over traditional paddy types has become a cause for concern among the experts. A news item in an Indian English Daily (*Indian Express*, 3 August,1998) quoting Dr. Sailabala Padhi, a botanist, says:

> the rapid replacement of locally-adopted seed varieties by high-yielding hybrid ones in contiguous areas may lead to the spread of diseases that can destroy the entire produce.

Further on, it adds:

> the practice might lead to situations like the Irish potato famine of 1845 and the Bengal rice famine of 1942.

The rapid depletion (extinction) of the rich traditional gene pool in Juangpirh is surely a pointer in that direction. In 1998, when the poor monsoon rains had ruined the crops in the plains, the hill slopes looked greener and the swiddeners expected a good harvest. Should we then think about food or protection of the forest first? For a poor country like India, this is surely a moot point.

The Juang are very much residents of Juangpirh in the hilly-forests. Recently, a housing colony by the side of National Highway No. 6 aimed to rehabilitate a group of Juang away from their traditional houses, and did not succeed because the Juang felt ill at ease in their new surrounding that lacked a forest setting. In fact, such schemes have been unfavourable with many forest-dwelling communities, destroying the strong bond between the indigenous man and his life-support-system, a point vividly highlighted by Ratha (1993). Forests need to be saved, and the Juang need to be brought to alternative food-getting avenues and, therefore, measures need to be taken. But should it be at the expense of an indigenous cultural system ? And would the environment (forest) remain the same when the culture that is so intimately attached to it, gets disturbed? Change is imperative and it comes inevitably to every culture, but it should be self-propelling and adaptive as Murdock (1971:320) has said. Progress should not be repulsive, leading to disintegration of indigenous cultures and, ultimately, eliminating natural resource management through cultural practices.

References

Dalton, E.T., 1872, *Descriptive Ethnology of Bengal*, Calcutta.

Das, G.N., 1964, 'Culture Contact and Social Engineering', *Adibasi*, 2:19-22.

Dash, J., 1988, 'Role of Kinship in the Social Life of Juangs of Kendujhar District', Orissa, *Adibasi*, 18(2):22-28.

Elwin,V., 1948, 'Notes on the Juangs', *Man in India*, 28 (1-2): 1-143.

Gupta, S.P., 1974, *Tribes of Chotanagpur Plateau (an ethno-nutritional and pharmacological cross-section)*, Ranchi : Govt. of Bihar.

Hunter, W.W., 1877, *A Statistical Account of Bengal*, vol. 19. London.

Joseph, S.J., 1994, 'Traditional Knowledge of Rural and Tribal People and Recognition of their Intellectual Property Rights, MSSRF approach', In: V. Shiva (ed.), *Biodiversity Conservation, Whose Resources? Whose Knowledge*? New Delhi: INTACH.

Kapp, W.W., 1986, *Social Cost of Business Enterprises*, Bombay: Asia Publishing House (2nd ed.).

McDougal, C.W., 1963a, 'The Structure of the Hill Juang', Ph.D. Thesis, New Mexico:University of New Mexico.

McDougal, C.W., 1963b, 'The Social Structure of the Hill Juang', *Man in India*, 43(3):183-90.

Mohanty R.P., 1991, 'Deforestation in Eastern Ghats and its Bio-social Impact on the Tribal population of Orissa', *Adibasi*, 31 (3-4):45-52.

Murdock, G.P., 1971, 'How Culture Changes', In: H. L. Shapiro (ed.), *Man, Culture and Society*, pp. 319-32. New York: Oxford University Press.

Radcliffe-Brown, A.R., 1952, *Structure and Function in Primitive Society*, London: Routledge & Kegan Paul.

Ratha, S.N., 1993, 'Forest and Tribal Culture : An overview of the Man-Plant-Animal Relationship', *Adibasi*, 33 (1): 1-4.

Richards, A.I., 1948, *Hunger and Work in a Savage Tribe*, Illinois: The Free Press.

Rout, S.P., 1969, 'Hand Book on the Juang', *Adibasi* 11 (1-2):1-97.

Singh, B. 1988, 'Forest Policy and Tribal Development', *Adibasi*, 28 (1) :5-9.

6

Economical and Biological Potential of Non-Timber Forest Products (NTFP) in the Similipal Hills of Orissa

Mihir Kumar Jena

NOWADAYS forestry research is oriented towards finding strategic solutions for sustainable development. It has become increasingly important to identify issues that focus on forests, forestry and sustainable development at the local level. The issues relating to Minor Forest Produce (MFP), now called Non-Timber Forest Produce (NTFP), is one such field which is oriented towards micro-level or local-level discourse. In many developing countries, including India, NTFP are central to the economies of many tribal and non-tribal rural households. Furthermore, dependence on NTFP is one aspect of the economic and ecological relationship of tribal and non-tribal forest dwellers within the forest. To meet the growing demand of NTFP, at least for the sustenance of a livelihood earning or domestic requirements, tribal and non-tribal communities have been exploiting forest resources since times immemorial. The competition for survival with income from NTFP in areas where forests have dwindled lead people to overexploit resources or to shift to occupational mobility. This usually happens in areas where forests have decreased due to the expansion of settlements, urbanisation, increasing population and market growth. Forest resources have been overexploited to meet rising demands, while commercial plantations and monocultures have reduced natural diversity, thus decreasing the availability of NTFP and therefore also their economic potential, in many areas of Orissa.

Thus, studying the biological and economical importance of NTFP, their marketing and conservation strategy in relation to the socio-economic life of forest-dependent communities is emphasised by forest and nature conservation policies and strategies to overcome unfavourable social living conditions.

Products obtained from forests were classified as 'minor', and 'major' when forests were valued mostly for commercial timber production (Jena 1996:198). In practice and fact, however, no forest produce is any longer minor from the social and economic point of view. NTFP, in the broadest sense, comprise commodities such as fuel, fodder, bamboo, cane, grasses, fibres, oil, tannin and dyes, gums, resins, medicinal plants, bark, leaves, flowers, fruits, tubers, mushrooms, seeds, mammals, birds, reptiles, fish and insects (Tewari and Campbell 1995). Traditionally, NTFP collection is a routine affair in the day-to-day life of forest-dwelling communities. For hunter-gatherer tribes, NTFP collection is considered a major source of life support. For tribes dependent on agriculture, NTFP is a source of additional income. The collection of NTFP by tribal communities was done traditonally to meet the requirements of local households. As a few commodities acquired more and more commercial value, reflected in a nationwide demand exceeding the local market perspective, trade and large-scale exploitation gained momentum.

NTFP policy

In the context of rural areas, NTFP management has gained importance for two main reasons. On the one hand, it aims at securing the livelihood of populations who live on the margin of a meagre subsistence economy and depend on NTFP. On the other hand, it is destined to maintain and upkeep the supply of raw materials for industrial production. The greatest worry in this regard is the depletion of forest resources and the loss of NTFP-providing species due to improper forest management. In India, a number of policies enacted in the past, such as the nationalisation of certain NTFP have had an ambiguous impact on NTFP-based rural livelihoods. Chambers *et al.* (1990) note that during the 1960s and 1970s when the state acquired monopoly rights over much NTFP the result was a reduction in the collection of NTFP and income for the tribal population. The National Forest Policy of 1988, while attempting to provide succour to forest-dependent communities as far as access to forest products was concerned, reflects the need to protect the rights and concessions enjoyed by tribals and the rural poor by providing them with priority entitlements to use forest products like fuelwood, fodder, NTFP and construction timber.

The report of the Committee on Forest and Tribals in India, published by the Ministry of Home Affairs, Government of India (1982), has stated very clearly that 'planning for tribal development or forest development in isolation has no meaning, the two must reinforce each other'. On the basis of the above understanding, the committee has recommended to:

(a) make a resource inventory survey of MFP.

(b) ensure the rights of the tribals to collect MFP which would be followed by the marketing of MFPs.

(c) accord priority on the co-operativisation of MFP in view of giving additional income to the tribals.

(d) open co-operative and private marketing channels parallel to each other in the interest of the tribals.

(e) assess the objective achievement of the tribal and forest development corporation.

(f) employ improved storage techniques using scientific methods.

(g) organise first-stage processing of MFP in tribal areas through co-operatives with the purpose of providing the tribals with a maximum benefit (Anonymous, 2001).

Highlighting NTFP as an important source of forest revenue, the Forest and Environment Department, Government of Orissa (2000), has stated the Subject-Policy on Procurement and Trade of NTFP. The gatherers of NTFP, the majority of whom are women, have limited access to the market and their primary concern is being able to dispose of the produce and getting fair prices for the NTFP. The prevailing system does not adequately provide these facilities to the primary gatherers.

Furthermore, the policy has empowered the Gram Panchayat / Gram Sabha in the scheduled castes and tribes and the non-scheduled areas to regulate the purchase, procurement and trading of NTFP in accordance with the guidelines of the policy (Anonymous p. 2). It reads that the Gram Panchayat/Gram Sabha in the scheduled areas will have the ownership over MFP within their territorial jurisidiction, (i.e. in respect of the MFP produced in and collected from government-owned land and forest land within the limits of revenue villages comprising the Gram Panchayat). According to the law, ownership of MFP in non-scheduled areas is not vested in Gram Panchayats. Gram Panchayats both in the scheduled and non-scheduled areas will, however, have the authority to regulate purchase, procurement (as distinct from collection by primary gatherers) and trading in accordance with the policy (*ibid*.,). However, such rights have been restricted on the part of Gram Panchayats in life sanctuaries and National Park areas.

In the policy guidelines, 60 NTFP items have been listed, 9 NTFP items are specially categorised for revenue purposes in consideration of their higher economic importance, and 7 varieties have been allowed to be procured and traded by the dealers who have been registered for this purpose. With the decentralisation of forest policies and the empowerment of local communities through JFM/CFM, significant changes have occurred in forest management. This has put people in a position to care for their forest resources in a sustainable way. The villages who could take control over good forests got better benefit, while villages who had scanty forest lost their access to many NTFP. This made people enter into reserve forests with the intention to collect as much NTFP as possible within a short time.

Micro-level studies conducted in various parts of the country reveal the multifaceted importance of NTFP for forest-dwelling communities. In this context, NTFP has become a major issue in Orissa and it is important to study the biological and economic potential of NTFP in different areas to understand the role of NTFP in tribal economy and the strategies of local diversification. The present paper focuses on a local assessment of NTFP in three different areas of the Similipal hills in Orissa. The areas studied are beyond the boundary of Similipal Biosphere Reserve and comprises six villages documented by resource inventories to understand the distribution, use and management of NTFP in different areas.

Similipal Hills

Similipal is located at the northern tip of the Eastern Ghats including eight undivided districts of Orissa lying between 21036' to 220 latitude north and 3605' to 36032' longitude east in Mayurbhanj district, comprising an area of 2750 sq km of forests. The Similipal National Park was established in 1956 with the objective of conserving the natural fauna, by preventing the wanton destruction of wild life and to provide recreation to tourists (Senapati & Sahoo 1976; Jena 1996). The ecological and economic prosperity of Orissa depends largely upon the preservation and conservation of the Similipal Biosphere Reserve. Similipal is an oval-shaped highland and a watershed with perennial water containing rivers like Budhabalanga, Sone, Gangadhara, Kharkhai, Bandhavi and Salandi fountains and spectacular waterfalls (Bose 1985, Jena 1996:54). It is an area of transition between the northern and southern flora of the country. There are animals from the Himalayas as well as from the plains south.

Although the Similipal hills are widely known as a compact dense forest, it bears about a century-old history of ethnic settlement and migration. There are ethnic groups like Hill Kharia, Mankidia, Bathudi, Ho, Santhal, Munda, Bhumij, Kolha, Juang and Lodha who are forest-based people living in and around the the Similipal hills.

The Similipal hill range is stocked with northern tropical moist deciduous forest, northern tropical semi-evergreen forest, mixed deciduous forests, high-level *sal*, dry deciduous *sal* forest and grass lands. *Sal* is the most dominant tree species and makes up an important part of the climax vegetation. There are 1076 species of land plants and 92 species of orchids. Amongst the plants, there are 8 extinct species, 8 endangered species, and 34 rare species. Amongst the animals, there are 42 species of mammals, 242 species of reptiles, and 12 species of amphibians. A census of wild animals indicates 99 royal bengal tigers, 119 leopards, 452 elephants, 850-950 bisons, about 10,000 wild boars, 9000 sambars, 2500-3500 deer, 4000-5000 antelopes and 1500-1800 mouse deers (Source: Director, Similipal Tiger reserve, Baripada).

Forest produce in Similipal

The Similipal hills, part of which are protected as a biosphere reserve, are a potential base for a number of NTFP and the availability of diverse NTFP in Similipal is undoubtedly better than in any other part of the state. Local people know about 80 different kinds of NTFP available, but there are only about 15 items in which the market is highly valued. The precious collections from Similipal include honey, arrowroot, dust resin, various kinds of gums, silk moth, lac, important medicinal plants, *sal* leaves and *sal* seeds, *mahua* flowers and various other seeds. Similipal is well-known for the production of honey and tribal people take this as an indicator of a good forest (Jena *et al.* 1997:80). Four types of honeybees have been identified (Jena 1996:200). The Hill Kharia, a hunter-gatherer tribe, are well-known for their expertise in honey collection, although other ethnic groups are also collecting honey. The Hill Kharia are also experts in resin and gum collection, arrowroot collection and processing and they collect the largest quantity of such products for the market. The Mankidia mainly collect Siali bark fibre, jungle jute, mat reed, from the reserve forests. Apart from these two communities, there are many other ethnic groups living in and around Similipal Biophere Reserve. These ethnic groups depend upon NTFP to supplement their income from other sources. Competition for NTFP collection is present among communities living outside the biosphere reserve because the market is closer to them as compared to the Hill Kharia and Mankidia. Further, with the proliferation of markets to procure diverse NTFP, the communities beyond the reserve area put more pressure on the forests and tend to overexploit these resources such as Santhal, Kolha, Munda, Bathudi and Bhumija. The general impression drawn from their perceptions envisage that the surrounding forests are still in a condition of providing or producing NTFP items in abundance.

First-hand field observations show that there are villages from where people visit distant forests to collect NTFP, requiring long walks up to 15 to 20 km to collect easily marketable NTFP like *sal* leaves. This puts tremendous pressure on resources when people from many villages depend on the same reserve as in the case of the forests around Bangiriposi. People from about 20-25 villages go there to collect *sal* leaves. The present study was directed towards identifying resource-rich zones, verifying standards of exploitation or collection, listing the main items exploited, and assessing the extent of pressure on the forest.

Methodology

Badampahar, Bisoi and Jashipur were sampled for making an assessment of the economic and biological potential of NTFP. The selection was done in consultation with an NGO (MASS) working at the district level and intensively promoting community forest management programmes in 718 villages of Mayurbhanj district. MASS has constitued area-wise federations of CFM committees in the district. A number of village-level CFM commitees constitute a cluster-level federation and a number of such federations constitute an area-level federation of CFM committees. In Badampahar, Bisoi and Jashipur, the success of CFM programmes and the supply of NTFP resources were assessed. Under each area, two villages were chosen in consideration of the number of households, resource availability, and the extent of dependence on NTFP, the quantity of collected forest produce, and the marketing facilities. Furthermore, the village selection was made in discussions with members of CFM committees, the area-level federation, and functionaries of MASS. Attention was given to select villages located outside the sanctuary area, for the study intended to measure the potential status of NTFP outside the biosphere reserve area. Out of the two villages in each area, one village had access to the biosphere reserve and the other village did not.

The study was conducted in May-June 1998 before the onset of the monsoon in the area. A rapid ethnobotanical appraisal was carried out to list the available NTFP items in the selected villages. Further data were collected in a specific format to compile resource inventories of those villages. To make an approximate estimation of the quantity of NTFP collected in a year, and to estimate the approximate price of NTFP items as per local market standards, data were collected through group discussions in which CFM committee members, women, and the village elite, participated. Simple statistical methods were used in the estimation and assessment of the price and quantity of NTFP.

Using the participant observation method, the assessment of the NTFP potential and making of resource inventories was quite fruitful because the participant observer was a person from the same community and was knowledgeable about the collection, processing, marketing, labour force participation and participation in JFM/CFM. However, the total collection of NTFP in relation to the forest area could not be calculated and compared because the local forests under control of the village-level CFM committees were not demarcated in segments. Local people collect the seasonally-available NTFP from open forests. The forests that are under the control of CFM committees can only be exploited by the villagers who come under the CFM committees. The available data, although measured and valued in approximate terms, were shared among the local people to find out major discrepancies. Discussions with community members and village forest committees, participant observation and PRA tools were employed to analyse the sustainability of NTFP items in the local forests. The research also attempted to understand what NTFP items are prioritised by local gatherers, and what role policy and market have in influencing such prioritisation.

Potential of NTFP in the study area

The result of the enquiry into the NTFP potential in the Badampahar area is presented in Table 1. As far as practicable, it has been tried to quantify the collections of NTFP items available in various seasons of a year and draw up a price index. Data collected in the villages of Dudhijharan and Badamchauradi envisage that people should have fair access to reserve forests and hence to diverse NTFP items. While the people in Dudhijharan are collecting as many as 40 marketable NTFP items, the people in Badamchauradi are restricted to only 12 NTFP items of commercial importance. In Dudhijharan village, a remarkable income is fetched from NTFP items like *sal* seeds, *mahua* flowers, black berry, wooden apple, Tamarind and Kalibahu (?); while in Badamchauradi the only notable income is obtained from the sale of *mahua* flowers. Further, the population in Dudhijharan village is far less than that of Badamchauradi village. From a calculation of the approximate price of the NTFP items it is deduced that the annual average income of a household in Dudhijharan is approximately, Rs.9,500 while in Badumchauradi it is only Rs. 660. There is a much higher labour input to collect NTFP in Badamchauradi as compared to Dudhijharan, as the availability of NTFP in their local forest is very limited. This implies that the potential income of a household from NTFP is dependent on biological diversity of the species. Apart from the collection of the said 40 NTFP items having market importance, there are many other items collected for domestic needs and livestock requirements. Such items are considered useful but less economically important, and therefore are not prioritised from an income point of view.

The results obtained from an investigation of NTFP in two villages of Bisoi area is presented in Table 2. The two villages studied in the area are Rajabasa with 52 households and Katas with 82 households. Primary gatherers of Rajabasa have access to the reserve forest, while the people in Katas do not. In Rajabasa, people collect 24 NTFP items mainly for sale in the market, while the people in Katas collect about 15 NTFP items for sale. In both villages, most of the income derives from the sale of *sal* leaves, *sal* seeds, *mahua* flowers, mushrooms and mango. Although the labour force input in NTFP collection is less in the case of Rajabasa, the villagers make a reasonable collection of NTFP from the local forests and reserve forests that have a good supply of NTFP resources. Accordingly, the average annual income per household in Rajabasa is calculated to be fairly high compared to that in Katas. A comparative look at the potential average income per annum from NTFP per household amounts to Rs. 33,000 in Rajabasa and only Rs. 3300 in Katas. The reason for this sharp contrast in income between the two villages is due to the remarkable variation in *sal* leaf collection. This difference stands because the villagers, particularly the women of Rajabasa village go to distant forests to collect *sal* leaves while the women of Katas do not. The women of Rajabasa spend four days a week collecting *sal* leaves in the vicinity of Similipal Biosphere Reserve. The rest of the week is spent in processing and stitching leaf plates. It has become a regular routine to exploit *sal* leaves from the biosphere reserve as it does not come under any CFM committee's control. In the local market there is a good demand for *sal* leaves. This is the main reason why people prioritise the exploitation of *sal* leaves even from the biosphere reserve area. Adding to that, *sal* leaves are available throughout the year and there is always a ready market for it. In these two villages women constitute the main labour force engaged in the collection of NTFP and this makes a major contribution to household income.

The status of NTFP in the Jashipur area was assessed by two studies in Begunia, having 73 households with better access to the reserve forest, and the other in Kankadani with 53 households. The Jashipur area is located at the outskirts of Similipal Biosphere Reserve and is a major marketplace for NTFP. The forest area in the locality is being gradually squeezed by the proliferation of the market and the expansion of settlements. However, the two villages considered here have access to the local forest, as well as to the biosphere reserve. The Forest Department closely monitors the collection of the villagers. The status of NTFP in the said villages presented in Table 3 indicates

Table 1: NTFP Potential in Badampahar Area

Sl. No	NTFP Item	Village : Dudhijharan Total households : 32		Village : Badamchauradi Total households : 88	
		Quantity collected	*Local market price in Rs.*	*Quantity collected*	*Local market price in Rs.*
1.	*Sal* leaves	200000 Nos.	2,000	1,42,000 Nos.	1420
2.	*Sal* seeds	150 Qtl.	30,000	-	-
3.	*Mahua* flowers	500 Q1tl.	24,000	58 Qtl.	27,840
4.	*Mahua* seed (Tola)	5 Qtl.	3,000	5 Qtl.	3,000
5.	Ebony Fruits	640 Kg.	32,000	6 Qtl.	3,000
6.	Char (*Buchanania lanzan*) seeds	32 Kg.	4,000	1 Qtl.	12,500
7.	Harida (*Terminalia chebula*)	20 Qtl.	2,000	-	-
8.	Bahada (*Terminalia bellirica*)	20 Qtl.	1,000	-	-
9.	Amla (*Emblica officinalis*)	10 Qtl.	2,000	40 Kg.	80
10.	Resin (*Jhuna*)	-	-	10 Kg.	-
11.	Marking nut	10 Qtl.	1,000	-	-
12.	Kusum (*Schleichera oleosa*) seed	25 Qtl.	12,000	5 Qtl.	2,400

Sl. No	NTFP Item	Village : Dudhijharan Total households : 32		Village : Badamchauradi Total households : 88	
		Quantity collected	Local market price in Rs.	Quantity collected	Local market price in Rs.
13.	Black Berry (*Syzigium cuminii*)	50 Qtl.	45,000	-	-
14.	*Siali* Seeds	10 Qtl.	6,000	-	-
15.	Mango	50 Qtl.	10,000	5 Qtl.	1,000
16.	Ghurudu (*Gardenia latifolia*) Fruit	40 Qtl.	24,000	-	-
17.	Pada (*Stereospermum suaveolens*)	15 Qtl.	-	-	-
18.	Deuncha (?)	15 Qtl.	5,000	-	-
19.	Ambada (*Spondias mangifera*)	5 Qtl.	200	3 Qtl.	120
20.	Jackfruit (*Artocarpus heterophyllus*)	10 Qtl.	1,500	2 Qtl.	300
21.	Custard apple (*Annona squamosa*)	10 Qtl.	1,250	-	-
22.	Wood apple (*Aegle marmelos*)	100 Qtl.	20,000	-	-
23.	Wild orange	2 Qtl.	3,000	-	-
24.	Karanja (*Pongamia pinnata*)	2 Qtl.	1,200	-	-
25.	Bamboo	100 Pcs	-	-	-
26.	Ant egg (?)	1 Qtl.	750	-	-
27.	Kali (Insect)	1 Qtl.	750	-	-

Sl. No	NTFP Item	Village : Dudhijharan Total households : 32		Village : Badamchauradi Total households : 88	
		Quantity collected	Local market price in Rs.	Quantity collected	Local market price in Rs.
28.	Neem (*Azadirachta indica*)	5 Qtl.	-	-	-
29.	Tamarind (*Tamarindus indica*)	100 Qtl.	37,500	5 Qtl.	1,875
30.	Phenphena (*Oroxylum indicum*)	1 Qtl.	500	-	-
31.	Sunari (*Cassia fistula*)	10Qtl.	1,500	-	-
32.	Amatha Gum (?)	2 Qtl.	-	-	-
33.	Yam	93 Qtl.	14,000	30 Kg.	4,500
34.	Malakendu (*Diospyros embryopteris*)	15 Qtl.	4,500	-	-
35.	Satabari (*Asparagus racemosus*)	15 Qtl.	15,000	-	-
36.	Patalgaruda (*Rauwolfia serpentina*)	2 Qtl.	1,500	-	-
37.	Bana kunduri (?)	5 Qtl.	1,000	-	-
38.	Bana kankada (?)	2 Qtl.	3,200	-	-
39.	Sikhat (?)	4 Qtl.	1,600	-	-
40.	Leafy vegetable	368 Kg.	-	50 Kg.	-
41.	Kali bahu	20 Qtl.	40,000	-	-
	Total		**3,01,550**		**58,035**

Table 2: NTFP Potential in Bisoi Area

Sl. No	NTFP Item	Village : Rajabasa Total households : 52		Village : Katas Total households : 82	
		Quantity collected	Local market price in Rs.	Quantity collected	Local market price in Rs.
1.	*Sal* leaves	14,40,00,000	14,40,000	57,60,000	57,600
2.	*Sal* seeds	150 Qtl.	30,000	500 Qtl.	1,00,000
3.	*Mahua* flowers	50 Qtl.	25,000	64 Qtl.	32,000
4.	*Mahua* Seed (Tola)	10 Qtl.	6,000	8 Qtl.	4,800
5.	Ebony Fruits	20 Qtl.	10,000	-	-
6.	Char seeds	25 Kg.	3,125	-	-
7.	Harida	10 Qtl.	1,000	5 Qtl.	500
8.	Bahada	30 Qtl.	1,500	-	-
9.	Amla	1 Qtl.	200	3 Qtl.	600
10.	Kusum seed	10 Qtl.	6,000	18 Qtl.	10,800
11.	Black berry	10 Qtl.	10,000	-	-
12.	Mango	100 Qtl.	30,000	85 Qtl.	25,500
13.	Jackfruit	50 Qtl.	7,500	2 Qtl.	300

Sl. No	NTFP Item	Village : Rajabasa Total households : 52		Village : Katas Total households : 82	
		Quantity collected	Local market price in Rs.	Quantity collected	Local market price in Rs.
14.	Wood apple	200 Qtl.	40,000	2 Qtl.	400
15.	Tamarind	10 Qtl.	5,000	10 Qtl.	5,000
16.	Phenphena	-	-	2 Qtl.	1,000
17.	Yam	25 Qtl.	3,750	-	-
18.	Ambada	10 Qtl.	1,000	100 Qtl.	10,000
19.	Mala kendu	20 Qtl.	6,000	-	-
20.	Leafy vegetable	500 Kg.	1,600	680 Kg.	2,000
21.	Mushroom	95 Qtl.	78,000	32 Qtl.	26,000
22.	Semul (*Bombax malabaricum*) cotton	5 Qtl.	8,000	-	-
23.	Rai berry	30 Qtl.	-	80 Kg.	-
24.	Wild horsegram	20 Qtl.	14,000	-	-
25.	Broom (*Thysanolaena maxima*)	30 Kg.	180	-	-
	TOTAL		**17,27,675**		**2,76,500**

that the locals collect only a limited section of NTFP despite their diversity. The people in Begunia earn money from the sale of *mahua* flowers, char seeds, kusum seeds, mango and other items listed in Table 3. The people in Kankadani specialise in *sal* leaves, *sal* seeds, tamarind, and mushrooms. From an estimation of the NTFP collection in the two villages it is deduced that the average household income per annum from NTFP is Rs. 4700 in Begunia, while it is Rs. 2800 in Kankadani. The variation in the annual average household income from NTFP is not very significant in these two villages. Labour input in Begunia is more than that in Kankadani.

With reference to Tables 1-3 it is evident that the availability of NTFP in the local forests is the determining factor of the income potential from NTFP together with the extent of dependence of local people on NTFP. Where there is a good biological potential of NTFP, the extent of dependence is higher than the areas where the diversity is low. While collecting NTFP, local people give importance to those products that sell best in the market, and ignore other kinds of NTFP available in the area. In areas or villages where the population size is bigger in comparison to settlements where access to forests is limited, people tend to overexploit the available resources, thereby causing more harm to the local forests. In the course of the development of awareness towards CFM, people in some villages have closed their local forest in order to prevent exploitation. This is a reason for occupational mobility for income generation from non-forest sectors. An interesting finding is that in villages where people enjoy the least access to forest resources, be it because of CFM or unavailability of NTFP in local forests, or restrictions imposed by the Forest Department, collectors go for the petty trading of NTFP. Such an example is seen in the Bisoi area. A number of people in Katas are occupied in trading with the collection of NTFP from people in Rajabasa. Thus, the NTFP marketing scene is changing fast. Earlier, the collectors were going to traders. Now the traders are knocking on the door of the collectors.

As seen in the study, out of many kinds of NTFP available, only 9-10 items have a market importance. Such items include seeds and leaves of *sal* (*Shorea robusta*); flower and fruits of *mahua* (*Madhuka longifolia*); seed of kusum (*Schleichera oleosa*); char (*Buchanania lanzan*) seeds; myrobolans (*Terminalia chebula*, *Terminalia bellirica*, *Emblica officinalis*); Ebony (*Diospyros melanoxylon*) fruits;

Table 3: NTFP Potential in the Jashipur Area

		Village : Begunia Total households : 73		*Village : Kankadani Total households : 53*	
Sl. No.	*NTFP Item*	*Qty. Collected*	*Local market price in Rs.*	*Quantity collected*	*Local market price in Rs.*
1.	*Sal* leaves	200000 Nos.	2,000	40,00,000 Nos.	40,000
2.	*Sal* seeds	98 Qtl.	21,000	66 Qtl.	14,520
3.	*Mahua* flowers	218 Qtl.	1,09,000	-	-
4.	*Mahua* Seed (Tola)	10 Qtl.	7,500	-	-
5.	Ebony (*Diospyros*) Fruits	18 Qtl.	9,000	-	-
6.	Char seeds	150 Kg.	21,000	78 Kg.	10,920
7.	Harida	100 Qtl.	12,500	-	-
8.	Bahada	50 Qtl.	2,500	-	-
9.	Amla	60 Qtl.	6,000	-	-
10.	Marking nut (*Semecarpus anacardium*)	5 Qtl.	500	15 Qtl.	1,500
11.	Kusum seed	100 Qtl.	60,000	7 Qtl.	4,200
12.	Custard apple	20 Qtl.	4,000	15 Qtl.	3,000

13.	Wood apple	50 Qtl.	5,000	-	-
14.	Phenphena	5 Qtl.	2,500	-	-
15.	Tamarind	2 Qtl.	10,000	50 Qtl.	25,000
16.	Sunari	-	-	1 Qtl.	150
17.	Mushroom	15 Qtl.	12,000	50 Qtl.	40,000
18.	Tasar	500 Pcs	2,500	80 Pcs.	400
19.	Honey	75 Kg.	3,000	20 Kg.	800
20.	Karanja	-	-	2 Qtl.	1000
21.	Dhatuki (*Woodfordia fruticosa*) flower	-	-	50 Qtl.	2,500
22.	Mango	150 Qtl.	30,000	25 Qtl.	5,000
	Total		**3,42,500**		**1,48,990**

mushrooms and tubers. These are the species found in almost each locality studied.

In consideration of the biological diversity and potentiality of NTFP in the study area, it is seen that people in the Bisoi area collect a significant amount of items available in the local forest. In contrast, having access to a larger number of NTFP items, the villagers from Dudhijharan in Badampahar area were not able to make a big income. Further comparison between the two villages indicates that in Rajabasa, out of the total NTFP collected, about 80 per cent of the income comes from the sale of *sal* leaves. It can be said that the economic potential of the NTFP is linked with the biological potential of *sal* leaves more than with any other item. The collection of *sal* leaves and *sal* seeds is also dependent on distance factors. When people have to move to distant places, they give emphasis to the collection of materials that can be easily carried. People in Rajabasa therefore prioritise the collection of *sal* leaves over the collection of *sal* seeds. Similarly, where the local forests do no have adequate juvenile *sal* bushes, people insist on collecting *sal* seeds as it is difficult to collect leaves from tall *sal* trees.

The prioritisation of NTFP during collection is made with regard to market demand. Since there is a ready market for *sal* leaves, *sal* seeds, *mahua* flowers and seeds, *kusum* seeds, char seeds, phenphena bark and fruits, myrobolans, marking nut, medicinal plants like *satavari* and *patalgaruda*, local people prioritise the collection of these items.

There is also information on the potential availability of NTFP that are not collected in the three areas. Data in this regard are based upon an approximate calculation of the potential resources that could have been exploited but are spared. It is observed that in Dudhijharan, people are not collecting dust resin and root of Ranu (*Cissampelos pareira*) that could have fetched Rs. 30,000. Similarly, in Rajabasa many kinds of NTFP are not collected which could generate income up to Rs. 55,000. They include *satavari*, Phenphena bark, leaves of *siali*, *patalgaruda* roots, Nageswar flowers, wild basil, Baidanka seeds and flowers of *dhatuki*.

Over the years, the forests were depleted very fast due to over-exploitation by the growing population and due to a lack of local management. With the JFM/CFM programme in the area, the local forests can be saved from further degradation. Gradually the forests have regenerated and biomass increased after local-level management practice was enacted with principles on the accessibility of forests, limits of exploitation and equitable solutions. Local NGOs have also contributed to the trend in the field of participatory forest management. However, the local-level management only applies to forests that are under community control and not to biosphere reserves.

NTFP and livelihood

The findings from the study are complemented by an overall assessment of the interdependence of NTFP collection and livelihood for the local people. The following points emerged from the perception of the primary gatherers and community members in the study villages.

All the forest-dwelling tribal and non-tribal communities exhibit dependence on forests for livelihood earning or domestic requirements. The degree of dependence varies with the availability of species in the forests. Roughly a decade ago, about 40 varieties of NTFP were available throughout the area, but at present the local people have confined the NTFP sources to certain pockets following indiscriminate exploitation of resources. At present one would have difficulty, finding 15 varieties of NTFP in unprotected forests.

- The villages closer to the biosphere reserve enjoy better access for bulk collection of NTFP, and their income potential is higher than that of the villages located farther away from the biosphere reserve.
- In many places, local forest resources have been badly depleted. This makes people to go to distant places to collect some common types of NTFP. Women walk long distances to collect fuelwood for their homes.
- The CFM committees at the village level argue that they have minimised their material needs from forests in order to increase the biological potential of NTFP-generating species. They make the further point that CFM committees have regulated the collection of NTFP with regard to equitable shares and have increased the regeneration of forests.
- Where village CFM committees have closed the forests that

are under their protection for the regeneration of forests, the NTFP-dependent people lose their rights in these forests. This often leads to outmigration and occupational mobility.

- In certain areas, local people have started trading the NTFP collected by the neighbouring villagers (e.g. people in Katas have started trading with the collections of villagers from Rajabasa). In the way, the hierarchy of primary traders at the village level is expanding. Once in trade, these new primary traders motivate the primary gatherers to collect more and overexploit the available resources.

- Problems in equity distribution of NTFP are seen in villages with multi-ethnic and multi-caste populations. In such villages some households' dependence on NTFP is exclusive while other households are marginalised. When CFM committees emphasise social equity, they ignore the economic needs of poor NTFP-dependent households. These households argue that they must be given a bigger share compared to households having alternative income sources. This resentment is highlighted when the poorest households do not participate whole-heartedly in CFM.

- Considering the presence of a market for NTFP in every nook and corner of the district, there is ample scope for the establishment of forest-based enterprises at places where the biological potential is good but the economic potential is low.

- The commoditisation of forest produce, have made the primary collectors categorise the existing flora into more or less commercially interesting plants. Better care is therefore taken for the commercially more valuable plants.

- There is a tremendous pressure on certain medicinal plants encouraged by outside traders. Overexploitation of the bark of the *paja* (?) tree, for instance, now threatens the species.

- *Mahua* trees are also threatened because the demand for charcoal made from *mahua* wood is increasing in the area.

- Certain tribal communities ignore the NTFP available in their locality which are collected by people from other ethnic groups. For example, in villages like Suanpal in the Jashipur area, the local Kolhs and Santhals do not collect *siali* leaves, whereas Bathudis walk a long way to collect *siali* leaves from distant forests. It is noteworthy that *siali* leaves are a lucrative NTFP that fetches a good price.

- The biological diversity of NTFP is not a direct determining factor of economic potential, as generally assumed. The economic potential of NTFP-dependent households is limited to a certain number of trusted NTFP like *sal* leaves, *sal* seeds, *mahua* flowers, *siali* leaves, and myrobolans.
- The Palli Sabha/Gram Sabha has not been able to check the collections from the Similipal Biosphere Reserve for two reasons. Firstly, the village committee has control over people as far as village forests are concerned. Secondly, the market moves to the forests to encourage the collection of raw, unprocessed forest produce. Further, the village committee has not been able to devise a monitoring system for NTFP collection and sale due to a lack of promotional efforts by the Forest Department.
- Local people are of the opinion that the forest-based co-operatives run by the State government do not give the required attention to procure the enlisted kind of NTFP. Bulk collection therefore goes to the hands of private traders at a lower price. To survive, local people are tempted to over-exploit resources.
- Resources are degrading in already depleted forests that are not (or only partly) under JFM/CFM management. On finding a market for every kind of NTFP items, irrespective of the price, local people will ultimately, overexploit the resources.

From the procurement status, the forest co-operatives must be able to identify resource-rich zones and must conduct resource inventories in such areas. Further, the forest co-operatives must enter into policy dialogues to express their views on how the resources in degraded forests can be regenerated or how resource use from degraded forests can be managed sustainably.

Conclusion

Looking at the NTFP scenario in the Similipal hills it has been suggested by local people that a community-level NTFP collection centre should be opened at each cluster of villages for a better marketing and monitoring of collection. Further, the community institutions should be given the authority to declare a forest open or closed to enable the monitoring of the biological and economic potential of NTFP diversity in every locality. The community institutions must be empowered to decide what NTFP items will be collected from local forests and what items must not be collected within a definite time frame for the propagation of threatened species.

An intensive need is expressed towards identifying resource-rich areas from the point of view of NTFP availability, and their management responsibility should be given to local communities. It is believed that community management of forests can improve forest conditions and increase the area of forests. To realise the fruits of the community initiatives through CFM, an NTFP-monitoring system needs to be devised at the community level for a continuous assessment of the biological and economic potential of certain NTFP.

Like the socio-economic environment, the role of forest resources is changing in a way that demands compatible management strategies and plans. The changing environment has caused local communities to change their perception of forest and NTFP resources along the lines of community forest management ideas. The role of forest resources has changed from the traditional view that forest resources are primarily meant to meet the demand for firewood, food and fodder, to being increasingly perceived as a source used for other purposes. This perception has been influenced largely by the heavy demand for forest products with high economic returns, it and by inappropriate natural resource use strategies. The extraction of firewood in the locality will increasingly become an issue for sustainable management in areas where forests are already in a degraded state. Further, to minimise local people's dependence on forests for livelihood earning, appropriate alternative income support measures at the village level must be broadly explored.

References

Anonymous, 2001, Impact of MFP items on Tribal Economy with Special Reference to Collection, Processing and Marketing : A Comparative Study in K.B.K. and Kondhmal Districts of Orissa, Report Submitted to Ministry of Tribal Affairs, GoI, by Council for Tribal and Rural Development (CTRD), Bhubaneswar.

Chambers R, N.C. Saxena, T. Shah, 1990, *To the Hands of the Poor: Water & Trees*, Oxford & IBH, New Delhi.

Government of Orissa, 2000, Policy on Procurement and Trade of NTFP, Resolution No. 5503/F & E, 31.3.2000.

Jena, M.K., Seeland, K., Patnaik, K.K.,1997, 'Where Trees Do Matter For Society', in Seeland, K. (ed.), *Nature is Culture*, pp. 79-89, Intermediate Technology Publications, London.

Jena, M.K., 1996, A study of Ethnobotany in Relation to the Social and Cultural Life of Certain Selected Tribes of Orissa, Ph.D Dissertation, Utkal University, Orissa.

Senapati, Sahoo, 1976, *Orissa District Gazetteer, Mayurbhanj*, Govternment of Orissa, Bhubaneswar.

Tewari, D.D. and J.Y. Campbell, 1995, 'Developing and Sustaining Non Timber Forest Products : Policy Issues and Concerns with Special Reference to India', *Journal of World Forest Resources Management,* vol.7, pp.151-78.

7

Forest Use among Some Tribes of Orissa

Kamala Kumari Patnaik

Introduction

FORESTS occupy the central position in a tribal economy. Tribal life, economy and society are profoundly affected by whatever happens to the forest. Forest represents for them a whole way of life; a home; a culture; worship; food; employment and income. Their folklore is full of references to the forest (Hivale/ Elwin 1935). There are several rites and rituals where forest produce is used in one way or other. Many of the leaves, climbers, and other items needed in magico-religious rites are collected from the forests. Many trees are considered to be the abode of deities and are worshipped as such.

The benefits to tribals from forests are various and are derived in a variety of ways which emphasise the social inclusiveness and cohesion drawn from their forest use and management practices. In areas where forests exist, not only the able bodied workers but also the old, infirm and children fetch something from the forests. It has been found that tribal people are not only familiar with a number of species around them, but also possess some understanding of the ecological interrelationship of various components of the forest. The appropriate use of the knowledge of tribals in forestry operations and the collection of MFP (NTFP) may add a significant dimension to development activities in the field.

Tribal forest management centres around two major issues — its sustainable use and preservation. These can be viewed at two levels: the macro level, involving a global distribution and use of resources, and the micro level, involving the dynamics of interaction between the forest dwellers and their immediate environment. The tribal exploitation of the forest is at the micro level and they consider that the forest provides livelihood-sustaining resources. They see the ecology and economy of production and adapt their own technology in exploiting the resources. Their ideas on forest ecology and specific resources make them to find new ways for collection and utilisation. Forests are always areas of multiple use for them. However, the days of self-sufficiency within their rural subsistence economy seem to be almost over with respect to many forest products. Mass collection of MFP for cash and as a supply service, (e.g. to the big tobacco, soap and pharmaceutical industries) has posed a serious challenge to sustainable forest management.

Forest use as a tribal lifestyle

The tribal forest-dwellers still depend largely on the forest for their economic survival and are occasionally blamed for the destruction of the forest. In the absence of a regular remunerative occupation, they have few alternatives other than to market forest resources for their survival. Low productivity in agriculture is partly compensated for by products from natural forests (see Table 1). The tribal forest dwellers are well acquainted with the biology of the forest, not in any structured form, but through experience. During the lean period they depend mainly on forest produce like *mahua* (*Madhuca longifolia*), *sal* (*Shorea robusta*) seeds, bamboo shoots, mushrooms, ebony fruits, different kinds of wild tubers, and leafy vegetables. They store dried leaves and flowers fruits and to be used as food when they are unavailable in the forest. Minor forest produce like myrobolans, broom sticks, arrowroot, resin and lac, are collected by the tribals from the forest. Poles, bamboos, creepers, and bark fibres are used for constructing houses and sheds. Ropes made out of the bark of trees and creepers are used to make cots and mats. Palm leaves are used to make mats and different types of grasses are used as brooms. Agricultural implements such as ploughs and fishing nets, as well as bows and other tools used for hunting, are taken from the forest.

Table 1: Important minor forest produce

Botanical name and family	***Parts collected***	***Collection time***
Acacia catechu (*L.f.*) *Willd.* (*Mimosaceae*)	Gum secreted from the trunk	Round the year, mostly in hot season
Asparagus racemosus Willd. (*Liliaceae*)	Tubers	Round the year
Azadirachta indica A. Juss. (*Meliaceae*)	Seeds	Summer
Bauhinia vahlii Wt.&Arn. (*Caesalpiniaceae*)	Leaves, bark	Mostly during Summer
Bauhinia variegata L. (*Caesalpiniaceae*)	Gum, bark	Summer
Buchanania lanzan, Spreng. (*Anacardiaceae*)	Kernel	Summer
Butea monosperma (Lam.) Taub. (*Papilionaceae/ Fabaceae*)	Leaves, seeds, flowers	Summer (seed, flower) Leaves all the year
Cassia fistula L. (*Caesalpiniaceae*)	Fruit, bark	Round the year
Celastrus paniculatus Willd. (*Celastraceae*)	Seeds	Late winter
Cissampelos pareira L. Var. *Hirusta* DC. (*Menispermaceae*)	Roots	Round the year
Curcuma zedoaria Rosc. (*Zingiberaceae*)	Rhizome	Summer
Diospyros melanoxylon Roxb. (*Ebenaceae*)	Leaves, fruit	Summer
Embelia robusta Roxb. (*Myrsinaceae*)	Seeds	Summer
Gardenia gummifera L.f. (*Rubiaceae*)	Resin gum	Summer
Helicteres isora L. (*Sterculiaceae*)	Bark fibres	Round the year
Holarrhena antidysenterica (Roth.) Wall. Ex DC. (*Apocynaceae*)	Bark, seeds	Seeds collected in summer
Jatropha curcas L. (*Euphorbiaceae*)	Seeds	Summer

Litsea sebifera Pers. (*Lauraceae*)	Seeds, bark	Summer
Madhuca longifolia Koenig. (*Sapotaceae*)	Flowers, seeds	Early Summer
Phoenix acaulis (*Palmae/Arecaceae*)	Leaves	Round the year
Phoenix sylvestris (L.) Roxb. (*Palmae/Arecaceae*)	Leaves	Round the year
Pollinidium angustifolium, Comb.nov. (*Ischaemum angustifolium*) Hack. (*Poaceae*)	Grass	Summer and Winter
Pongamia pinnata L. (*Papilionaceae/Fabaceae*)	Sticks, seed	Seeds during summer
Rauvolfia serpentina Benth. Ex. Kurz. (*Apocynaceae*)	Root	Round the year
Schleichera oleosa (Lour) Oken. (*Sapindaceae*)	Fruits, seeds	Summer
Semecarpus anacardium L.f. (*Anacardiaceae*)	Marking nuts	Summer
Shorea robusta Gaertn.f. (*Dipterocarpaceae*)	Leaves, seeds, resin, tannin	Seeds in late summer
Terminalia chebula (Gaertn.) Retz. (*Combretaceae*)	Fruits	Early summer
Terminalia bellirica (Gaertn.) Roxb. (*Combretaceae*)	Fruits	Early summer
Terminalia arjuna (Roxb.ex. DC) W.&A. (*Combretaceae*)	Bark	Round the year
Tamarindus indica L. (*Caesalpiniaceae*)	Fruit, seeds	Winter and early summer
Thysanolaena maxima O. Ktze. (*Poaceae/Gramineae*)	Branches	Early summer
Woodfordia fruticosa Kurz. (*Lythraceae*)	Flowers, roots	-

SAL

One of the most prominent trees for tribal consumption is the *sal* tree. Normally *kendu* (*Diospyros melanoxylon*) leaves are used as tobacco wrappers to make domestic cigarettes. But *sal* leaves are also extensively used for this purpose. *Sal* twigs are used as brush sticks; the leaves are used to make cups and bowls; and the wood is a good quality fuel. The most important means of energy for cooking is dry wood collected from the forest. The demand for good quality fuelwood, such as young *sal* trees, is constantly rising, and except a few species like *Emblica officinalis, Strychnos nux vomica, Diospyros melanoxylon, Ficus* species *and Eucalyptus,* all other tree species are cut for fuelwood. *Sal* is the best wood for building houses and making agricultural implements. Dust resin is collected from *sal* trees which also host *tassar* silk worms. Another important aspect of the tree is that it is deified. The tribals worship in the *sal* groves (*Zaheera*) in the eastern and northern parts of Orissa.

MYROBOLANS

Fruits of *Terminalia chebula, Terminalia bellirica* and *Emblica officinalis* are widely collected in the tribal areas. While its collection and sale is intensive in districts like Mayurbhanj, Kendujhar, Koraput, Phulbani and Ganjam, in other areas their collection is less important. Myrobolans are spread throughout the forest tracts of the state. Amongst all three varieties of Myrobolans available, the *Bellirica myrobolan* and *Emblica myrobolan* are inferior to the *Chebulic myrobolan*. Myrobolans are mainly used as Ayurvedic medicine. The fruits are collected during the months of December and January, dried in the sun by spreading them evenly on the floor for a fortnight. All members of the family are engaged in this activity irrespective of sex and age. After collection, the drying is generally entrusted to the children, while playing at home. After drying, the fruits are sold in the market or by auction. Some merchants also come to the villages and purchase the trees or the raw fruits from the tribals at a very low price. Myrobolans do not need elaborate processing or special care for storage, and dried fruits can be kept for years without any deterioration or loss of quality. However, tribals cannot afford prolonged storage as the fruits fetch money quickly enabling them to buy food. Sometimes the price of myrobolans falls under a certain limit and agricultural labour becomes financially more attractive than gathering, which has a positive effect on the propagtion of these species.

MAHUA FLOWERS AND SEEDS

Mahua trees (*Madhuca longifolia*) grow wild throughout the tribal areas of Orissa and their flowers are collected by the tribals on a large scale. The flowers are used mainly as a raw material for the preparation of home-made liquor both by tribal and non-tribal people. It has a very important position in the tribal culture, both as an economic commodity, as well as an item associated with religion and folklore. Besides, the flower is also used for consumption during the lean season of the year. The *mahua* flower is collected during the early summer months between February and April. Generally women and children are engaged in the collection of *mahua* flowers and men occasionally visit the forest along with them when there is no other work for them to do. The flowers are brought home and dried in the hot sun by spreading it on mats in front of their houses. Well-dried flowers which turn dark brown from white can be stored to be sold for cash, or kept for self-consumption during the lean months. It is either used as gruel by boiling it with water or as a cake by frying it. Its sweet taste is relished by children as well as adults. *Mahua* flowers are also ground with *sal* seeds and a cake is prepared out of the paste. The flower is used predominantly for the preparation of *mahua* liquor which is very popular in the tribal areas. It is widely used in their social activities like festivals, marriage and death ceremonies. On the occasion of arranging a marriage, a quantity of liquor forms an important part of bargaining on the bride price and is served during the actual performance of the marriage ceremony. It is also used while propitiating the deities during the performance of various village festivals. Pure *mahua* liquor is given to mothers after child birth as a stimulating medicine. Further, pure *mahua* liquor is also used as antiseptic lotion applied on fresh cuts and injuries. A *mahua* tree is very economic to the tribals. From the flowers and fruits of a single mature tree a tribal family can easily earn an amount of approximately Rs.600 per season. *Mahua* seeds are harvested a month after the collection of *mahua* flowers during the months of June and July. *Mahua* seeds are also used for the production of edible oil.

TAMARIND

Tamarind is commonly collected either as green tamarind or as a mature fruit by the tribals. Tamarind is a semi-wild tree growing in the dense forest. It is also grown traditionally by tribal households around their houses and in southern Orissa one fails to find a village without a Tamarind tree. It occupies a prominent position in tribal life because of its utility. The Kondh (a Dravidian tribe) and Saoras (a Mundari-speaking tribe) consume the leaves at an early stage, and the fruits green or ripe. Wood is used as timber for the construction of houses, sheds for cattle, and for fuelwood. Furthermore, Tamarind commands a wide national and international market, which has made its collection and sale by tribal people increasingly important. Most of the tribal communities (Saora, Kuttia Kondh, Dongaria Kondh) lease out their individually-possessed tamarind trees for a fixed amount of money. Tamarind, when it is collected by tribal households, is sold either with the shell or after removing the seeds to private traders. Mostly, Tamarind seeds fetch more money as compared to uncrushed Tamarind shells, but tribals generally do not take the pain of removing the seeds and are satisfied with whatever price they get for the shells. The Tamarind season starts in January and ends in March.

MISCELLANEOUS FOREST PRODUCE

Lac is an important cash crop produced in the forest. The *kusum* tree (*Schleichera oleosa*) and *palas* (*Butea monosperma*) are the host plants of lac insects. The tribes of eastern Orissa are very experienced in lac production. But as most of the host plants have disappeared due to deforestation, the tribals have suffered a heavy loss in their economy.

Tassar sericulture is one of the most important sources of household economy of the tribes. The tribes of Orissa are well acquainted with the technology of *tassar* production. They have identified species like *sal* (*Shorea robusta*), *arjun* (*Terminalia arjuna*), *asan* (*Terminalia tomentosa*) as primary host plants of the silk moth. The tribals identify the larvae of silk moths and leave them on the leaves of these host plants which produce cocoons once in a year.

In the summer, *char* (*Buchanania lanzan*) seeds which fetch high prices in the market are sold to traders for a throw-away price. In Mayurbhanj District, the tribals barter *char* seeds for 4 kg of salt or similar items.

Siali leaves of *Bauhinia vahlii* is another important item collected on a large scale in the tribal areas of Orissa. They are used as plates by stitching a number of leaves together. Leaves are collected in two seasons, namely between May and July, and between November and January. The collection of leaves is not confined to any particular tribe. However, with the advent of JFM many forest areas are encroached

Table 2: Multiple use of habitats

Method of Production	*Sub-habitat*	*Product*	*Usage*
Agriculture	Foot hills, valleys, wide slopes	Paddy, Millet, Pulses, Oil Seeds.	Food,
		Paddy Straw	Fodder / Fuel
		Water	Drinking / Irrigation
Horticulture	Middle hills, steep slopes	Orange, Banana, Coconut, Cashew, Mango, Jackfruit, Pineapple,	Food,
		Banana	Detergent,
		Coconut Stem,	Fencing,
		Timber	Furniture and Fuelwood
Rearing livestock	Grazing in the	Dung,	Manure
		Milk,	
	forest, stall-feeding	Draught force,	Food
			Ploughing, Cart pulling
		Hide	Leather making
Gathering	Grassland	*Themda* sps. Grass	Thatching
	Scrubby jungle	Leaves (dry/fresh), Twigs, Branches, Timbers	Manure, Mulch, Fuel, Construction
	Hillocks with forests	Leaf (dry/fresh), Twigs, Branches, Grasses, Food Items, Timber, Fuelwood	Manure, Mulch, Grazing, Construction Cooking
	High quality	Honey, Wild Fruits,	Food, Medicine,
	forests	Flowers, Herbs,	Spices,

		Pepper, Soap, Nuts, Cane, Bamboo,	Detergent, Weaving Baskets, Mats, Furniture, Cash, Food
Apiculture	Forest flora	Honey	Food
Sericulture	Natural forest or *sal* plantation	Silk	Fibre
Fishing	Rivers/streams	Fish,	Food

upon by different villages or village associations. The tribals of each village will collect leaves from the forest surrounding their village, spreading over some kilometers and will not tresspass into the forest belonging to another village. The leaves are collected by women and children and dried in the sun for four or five days. After the leaf is sufficiently dry, it is stacked into bundles and sold. The Mankidia of Mayurbhanj district of Orissa make very good rain jackets from *siali* leaves, sown together. Properly dried leaves make the jackets last for one rainy season.

Ecology as a tribal way of life

HUNTERS AND GATHERERS

Tribals use their traditional skills and local knowledge in exploiting renewable forest resources. The life of the Mankidia, for instance, is very much associated with the forests. The Mankidia are a nomadic people who are believed to be the descendants of the Birhor tribe. Bir means forest, and Hor or Hol is human being, in their native language (Pathi 1993:24). Birhor are mainly of two types: Jaghi — the settled, and *Uthulu* — the nomadic. The original place of Jaghi and Uthulu Birhor is the Chhota Nagpur plateau of Bihar. They camp in the forest and collect their desired materials, mostly bark fibres of *siali* (*Bauhinia vahlii*), a creeper, from the forest. Leaf huts which are called Kumbha, and look like a dome or conical tent, are a characteristic feature of Mankidia forest life. The Kumbha is built on four to eight small bamboo or wooden poles of about one-and-half metres in height. The hut is mainly covered with leaves of *sal* (*Shorea robusta*), marking nut (*Semicarpus anacardium*), *teak* (*Tectona grandis*), *rai* (*Dillenia pentagyna*), *siali* (*Bauhinia vahlii*), and with other leaves when the above are scarce or not available.

The door called Bagdir is made of twigs and leaves and closed when the family members are out for hunting and gathering in the forest. They do not require any wood for the floor of the hut. If the camping site is not a clean place then a leaf bed is sometimes made. About 10 to 15 families camp in a place which is called Tenda. The Birhor camp mainly in and around the Similipal hill range and normally stay in one place for one to two months. They migrate with their lineage members on a date fixed by their elders. This fixing of dates is done on weekly market days where a number of groups meet and exchange their news and views and decide where to go next. When they decide upon a new place, they first consider the availability of forest resources there. In the rainy season, they make their Tenda close to the forest from where the market is easily accessible and sell their products. In the winter season, they shift their camp to two or three other places. In the summer season they mainly collect wild roots and tubers. The Mankidia consider the bark of *siali* (*Bauhinia vahlii*) their main forest produce. They collect it in the rainy season, because at that time it is much easier to peel off the tree, as it is wet and soft, while in the dry summer season the bark may break or crack when it is very hot and there is a risk of forest fires. The wanton destruction of the forest by tribal communities who practice shifting cultivation, and the non-tribal people who gather fuelwood from the forest, make it difficult for the Mankidia to get the required amount of bark and are forced to have change their camps continuously. The *siali* bark is processed into various types of ropes and containers such as the *sikur* or sling which is used to carry goods; the *pagha* which is used to tether cattles; and the *pandachhati* used to tether buffaloes. *Chhelipagha* is used to tether goats, and *dhaunri* for tying cattle; *barkai* is used for drawing water vessels from wells; *Tupa,* a small vessel, is used for pressing oil seeds. Flat barks are used in making *tupa* and cots. Thin strips are used in making nets, traps, tethering ropes in which small pieces of bark are also utilised. There is always a good demand for these products in the local markets. As far as the division of labour is concerned, Pathi (1993:21) states that the collection of siali fibre for making ropes is the primary occupation of Mankidia men, while hunting and collection of other forest resources are their secondary occupation. They are skilled monkey catchers for which they make nets for themselves which are never for sale. Household activities are the primary occupation of all adult females, and the gathering of forest products, rope-making and leaf-plate making, are their secondary occupation. Patnaik (1987:7) observed that except in the case of baskets and ropes used for tying buffaloes, which are made by men, all other types of ropes are made by both men and women. A head-load of *siali* bark is used up within two to three days.

As far as other forest products are concerned, they usually collect resin and silk-cocoons for sale, and fruits, roots, mushrooms, and leaves for their own consumption. Sometimes, the Mankidia go for group hunting (*sendra*) into the forest. During the expedition, they first place traps, snares and nets for birds, hares, mongoose and monkeys, and then hunt other game in the forest.

They have also started to collect honey by learning the technique from the Hill Kharia, a neighbouring tribe, and causing a considerable loss to the latter's economy. The Hill Kharia depend upon honey collection and sale which constitutes their major source of cash income. As traditional collectors of honey, they know certain species of plant which the honeybees usually forage upon. Among many species, the *siali* (*Bauhinia vahlii*) flowers is one of the most important sources of nectar for the honeybees. Patnaik & Jena (1995:10-11) report that the Hill Kharia observed that each 'state of honeybees' has their own territory to collect its nectar. The insects make use of all sorts of flowers in all tiers of the forest. Honeybees usually make their hives at a great height above the ground on rocks and in high trees. A good population of honeybees, or a good quantity of honey production, indicate a virgin forest flora. A reduction in the floristic diversity in the area would affect the honey production.

In dry months, honeybees collect nectar from plants like *Crataeva religiosa, Caesalpinia* species, *Terminalia alata, Aegle marmelos, Buchanania lanzan, Dichrostachys cinerea, Adina cordifolia, Sterculia urens, Syzygium cumini, Prosopis julifera, Soymida febrifuga, Grewia tiliaefolia, Shorea robusta, Bauhinia vahlii*. In the winter season, the species which are preferred by the honeybees are *Carum copticum, Guizotia abyssinica, Coriandrum sativum, Tridax procumbens, Ageratum conyzoides, Mangifera indica*. Although the Hill Kharia identify many species as important from the nectar production poin-of-view, *siali* (*Bauhinia vahlii*) flowers make up the major part of the total nectar collection during the summer season. When Mankidia slash *siali* creepers, the Hill Kharia loose a substantial quantity of honey.

The neighbouring tribes of the Hill Kharia namely Bathudi, Ho, Santal, Munda, Bhumija and Mankidia have taken up honey collection which inevitably leads to an over-exploitation of wild bee-hives, a decreasing availability of honey, and a declining source of income for honey collectors due to competition. A second important activity for the Hill Kharia is to collect *sal* resin that oozes out from the tree bark in a semi-liquid form during the rainy season, and is mainly found on the upper part of tall *sal* trees. According to Jena (1996:202):

> the people believe that sufficient rain and wind twist the upper part of the tree, which causes cracks on the tree trunk through which resin comes out. In case of little rain and wind, *sal* resin is not found in sufficient quantity. Besides, it is also believed that a particular type of insect makes holes or pores at several points on the upper part of the *sal* trees through which resin comes out.

Arrowroot is also commonly collected in the forests by various tribal people. The Hill Kharia collect wild arrowroot whereas the Mankidia do not. In the south, the Dongaria Kondh collect them but the neighbouring Kuttia Kondh and Lanjia Saora do not. The processing of arrowroot needs a series of labour-intensive processes for several days, as well as careful and constant watching. The Hill Kharia are able to identify wild arrowroots according to their suitability for further processing. Jena (1996: 202) mentions five varieties of arrowroot known to the Hill Kharia, of which the first three are considered to be the best: the *bhandura* (white) variety contains more water than the others, while the little arrowroot, *chauli* (white), contains less. The *haladia* (yellow) variety contains a medium quantity of water, as well as the *balunga* (white) and *kanka* (yellow) varieties. The Dongaria Kondh distinguish only between two varieties of arrowroot, namely white and yellow. The Desia Kondh have taken up the cultivation of white arrowroot. The Hill Kharia collect and process arrowroot to powder and sell it in market. The Dongaria Kondh barter the rhizomes for paddy.

SHIFTING CULTIVATION, PERMANENT DRYLAND AND IRRIGATION FARMING

After this short account on some aspects of forest use by the nomadic Mankidia and the semi-nomadic Hill Kharia, some north Orissan hunter-gatherer tribes, and groups of shifting cultivators in south Orissa will be looked at. The Kondh and Saora communities constitutes the major population among shifting cultivators. There are many sub-sections of Kondh, out of which the Dongaria Kondh and Kuttia Kondh are the so-called primitive communities. In the case of the Saora, the Lanjia Saora are considered to be the most primitive section of the tribe.

The Kuttia Kondh in the Belghar area of the newly-formed Kondhmal District of Orissa depend exclusively on shifting cultivation. The shifting cultivation includes a series of successive stages. Normally, the agricultural year begins in March-April with the slashing of the forests. A number of criteria are considered when a site for shifting cultivation is selected. The cultivation technology has undergone enormous changes in recent times. The Kuttia Kondh cultivate the hill slopes in a rotation cycle of about 3 to 5 years of fallow period, which was about 10 to 12 years only three decades ago. The Kuttia Kondh practice multi-cropping with about 25 species of cereals, pulses and vegetables in the first year, whilst in the second year they only cultivate some little millet (*Panicum miliare*) in the same fields.

The fallow period is only of minor importance to the the shifting cultivation practices of the Lanjia Saora and Dongaria Kondh of Rayagada District. Both communities practiced shifting cultivation exclusively in the past, but have now taken up permanent cultivation on terraces (Mahapatra, 1990:3). The Saora have turned from shifting cultivation to paddy cultivation on terraces which has resulted in a change in food habits and lifestyle. Emigration to Assam and Arunachal Pradesh, where they worked in the the tea plantations, resulted in innovations in their agriculture. Their newly-acquired skills in terrace- making and contour-bonding enabled them to irrigate their crops by channelling the hill streams. There are also dry terraces depending upon rainfall. With migration for labour income and the adoption of terrace farming they have neglected and partly abandoned shifting cultivation. To produce annual crops such as, pulses like cow pea and pigeon pea, maize and castor by dryland farming, has caused a severe denudation of the hill slopes. Decreases in yield due to a remarkable loss of fertility and soil erosion are evident. With the introduction of cashew nut cultivation as a permant crop, soil degradation could be stopped to a certain extent as cashew nuts are a good cash crop which might compensate a declining income in other spheres of their rural economy. Mahapatra (1990:3) emphasises the fact that swidden land is parcelled into individually-owned plots in south Orissa, which does not deter the owner from over-cultivating the land, thus leading to degradation as a consequence of privatisation.

In sharp contrast to the Kuttia Kondh and Lanjia Saora communities, the Dongaria Kondh are horticulturists. They have switched from shifting cultivation to horticulture due to the hereditary distribution of land, and an increase in population pressure that has led to a degradation of soil fertility. The climatic and soil conditions are favourable to the growth of fruit trees and orchards in the Dongaria Kondh area. They became experts in growing jackfruit, banana, mango, pineapple, citrus and turmeric for which the market is always good. But the Dongaria have not fully given up shifting cultivation. This is because of their food habits which favour locally-grown varieties such as little millet (*Panicum miliare*), finger millet (*Eleusine coracana*), and *sorghum* (*Sorghum vulgare)*. They have also taken up paddy cultivation in the plains, wherever available. Permanent agriculture on hill slopes and terraces have stimulated the Dongaria Kondh to identify different zones on a hill slope that favour different kinds of crops. Each part of the slope has a specific denomination which is an indicator of the diversity of space management. Their land use pattern and utilisation of natural resources has proved to be effective, economically viable and sustainable in the context of a partially market-oriented self-sufficiency.

Shifting cultivation has been converted into permanent agriculture in places like the Kashipur block of Rayagada District. There, the Desia Kondh who practice shifting cultivation on the hill slopes make up the major part of the population. The terrain in the Kashipur block is different from that of the Kuttia Kondh and Dongaria Kondh habitat. The slopes are fairly wide providing opportunity for ploughing. People have begun to cultivate paddy there permanently. The hills have been denuded of forest vegetation and in a few places terraces have been built. Shifting cultivation has become very rare appearing only where it is appropriate to the soil type, physiography and ecology of the hills, and where ploughing and permanent cultivation do not seem to be adequate. Elsewhere, swiddens have been completely converted into other forms of land use.

Some conclusions and suggestions

Although forests and forest produce are the primary resource base for the tribal communities of Orissa, some tribal communities adhere to distinct uses and particular management pratices. Changes in land use and property regimes have often led to innovations in lifestyle and strategies for survival. In the context of resource management, their skills have been adjusted to new social conditions and have thus contributed to a by-and-large sustainable management of renewable natural resources.

Owing to the scarcity of particular products, or to the total depletion of forests due to shifting cultivation, tribal communities have moved towards alternatives. They have reshaped their management practices to improve the resource base by horticulture (Dongaria Kondh); by terracing; by taking up the cultivation of new crops; by trying to improve the fertility of the hills by regular fallowing (Kuttia Kondh); or by introducing new agricultural techniques (plough cultivation by the Desia Kondh of Kashipur). These initiatives show new avenues for resource use, as well as a cultural flexibility and readiness to innovate with regard to agriculture and multiple use of forests. In the search for a new resource base where tradition and adaptation to modernity meet, indigenous forms of agro-forestry would deserve more encouragement by the land use and forest policy in Orissa. In this context, ethno-ecological research must be further developed to study the ecological and socio-cultural effects of such adaptational processes of agricultural land use within local forest management. This will be useful in view of the sustainability of natural resource regimes in general, not only in the tribal areas of Orissa.

References

Hivale, S., Elwin, V., 1935, *Songs of the Forest. The Folk Poetry of the Gonds*, Allen & Unwin, London.

Jena, M. K., 1996, A study of Ethnobotany in Relation to the Social and Cultural Life of Certain Selected Tribes of Orissa, Ph. D. Dissertation, Utkal University, Orissa.

Jena, M.K., Seeland, K. & Patnaik, K.K., 1997, 'Where Trees do matter for Society: The Socio-Cultural Aspects of Sal (*Shorea robusta*) and Salap (*Caryota urens*, L.) in the Similipal Hills of Orissa, India', In: Seeland, K. (ed.), *Nature is Culture*, Intermediate Technology Publications, London, pp. 79-89.

Mahapatra, L. K., 1990, 'Ecology and Technology of Livelihood and determinants of Tribal Economy in Eastern India', *Adibasi*, THRTI, Orissa, vol. XXX, no.283:1-4.

Pathi, P., 1993, 'Women and Development : A study among *Birhors* and *Santals*', M. Phil. Dissertation, P.G. Department of Anthropology, Utkal University, Orissa.

Patnaik, K.K., Jena, M.K., 1995, 'Conservation of bio-diversity : A tribal perception of Similipal biosphere reserve', In: R.C. Mohanty (ed.), *Environmental Change and its management*, Kamala Raj Enterprises, Delhi, pp. 9-12.

Patnaik, N., 1987, 'The Birhor' *Adibasi*, THRTI, Orissa, vol. XXVII, nos. 283:1-17.

8

Joint Forest Management

Merger of two Traditions?

Nadja Ottiger

Introduction

JOINT Forest Management (JFM) in India has gained ground since the 1980s as a promising strategy for the protection and conservation of local forests. My intention here is to draw attention to the major challenges to be overcome to make JFM a feasible strategy for forest conservation. For this purpose, I would like to discuss some basic features of the management systems of forest-dwelling communities and Forest Departments. I will then raise some questions concerning the feasibility of JFM strategies on the basis of their respective traditions of forest management. Such questions refer to practical aspects of forest management such as silvicultural techniques and practices; management objectives; needs and priorities; and finally, to institutional aspects of control over forest resources, and the question of legitimacy of a people-oriented approach in JFM strategies. I may anticipate that I do not have the answers to all my questions. The aim of my paper is to provoke suggestion for possible solutions and to indicate fields of further research. My questions are based mainly on the secondary literature dealing with the situation of forests and people in India, especially in Orissa, as well as some of my own field experiences.

The following characterisation of the traditions of the two principle forest management actors is rough and claims in no way to fully reflect the manifold realities of forest communities and Forest Departments in time and space of India. Instead of giving another individual case study, I would like to outline some general considerations which may point to further research to be carried out among forest-dwelling communities, and in Forest Departments.

What type of forest management?

Do forest-dwelling communities dispose of forest management systems?

To answer this question, any discussion depends on the definition of the concept 'management'. Against a narrow definition of forest management biased towards Western conceptions K.F. Wiersum advocates the following broad definition of forest management:

> Forest management should be considered to involve not only silvicultural practices, but all conscious human activities directed at maintaining its production capacity. It can best be defined as the process of making and implementing decisions about the use and maintenance of forest resources and the organisation of the related activities (Duerr *et al.* 1979). It refers to the total set of technical and social arrangements involved in the protection and maintenance of forest resources for specific purposes, and the harvesting and distribution of forest products. — 1997:7

This definition includes forest-use systems with a rather low degree of human manipulation, even practices without silvicultural management at all. Whereas other authors equate forest management with the conscious manipulation of vegetation to promote the maintenance and/or productivity of forest resources, (i.e. with silvicultural practices) Wiersum operates with the criteria of controlled forest extraction. In this way, if rates of extraction of forest products are deliberately regulated in order to maintain a forest's productivity, this is a matter of forest management (1997: 7). So even hunters and gatherers, collecting wild flora and hunting fauna in a regulated way, are not only users but managers of forests. Similarly, a sacred grove where the extraction of forest products is prohibited for the sake of forest-dwelling spirits? can be considered a managed forest patch. The access to the sacred grove is controlled with the specific purpose to fulfil the religious aspirations of the forest-dwelling community.

Controlled use and/or manipulation of forests for specific purposes without endangering the supply of the required forest resources may be considered to be necessary features of any forest management system. However, the degree of control and manipulation; the nature of purposes and needs; the 'total set of technical and social arrangements' involved; and the systems for the distribution of forest products vary widely.

Wiersum's definition implicitly encompasses a criteria of sustainability as he defines one aim of any systems 'to protect and maintain forest resources for specific purposes'. 'Reap and run' strategies[1] cannot therefore be considered as 'forest management systems'. Let us now consider some characteristics and principles of indigenous management in comparison with professional systems. I will start with the needs which drive both forest users as well as forest departments.

NEEDS

Local communities' needs for forest products tend to be manifold. Forests and forest products are used locally for a broad variety of socio-economic and cultural purposes: for fuel, food, fodder, manure, medicine, building material, grazing, providing material for agricultural tools and for handicrafts. In communities which practice shifting cultivation, forests are needed temporarily for agricultural purposes. Apart from such material needs, forest are used for social and religious purposes, as hunting grounds, burial places, sacred groves, and so on.[2] (The case studies of the Juang, Kuttia Kondh and Dongaria Kondh presented in this volume are illustrative in this respect). Forests supplying such a broad range of forest products and functions are ideally forests with a high degree of biodiversity. Tribal communities do not generally try to create a homogenous stock of trees. However, it cannot be concluded that indigenous communities consciously manage forests in order to maintain a high degree of biodiversity. Wiersum calls attention to the fact that:

> in many cases management practices are primarily directed at forest resources: those attributes of a forest which are considered for fulfilling human needs. These needs are often

---

1. Colombijin (1998) uses this term to refer to unsustainable forest resource use practices by Indonesian frontier societies in the nineteenth and twentieth century.
2. Fernandes *et al.* (1988) provide an excellent study on the demand for forest resources by tribal communities in Orissa.

> of material nature, but cultural and religious needs may also be involved. — 1997:7

The use of forest resources in forest-dwelling communities is embedded in an economic system which is still partially oriented towards subsistence. Many forest resources are used directly for maintaining the livelihood of the people. However, there is a trend away from using forest products for self-consumption to selling forest resources. Pressure on forest-dwelling communities to earn some monetary income is growing and local forest-dwelling communities may also be dependent on forest resources for cash income, from local markets. In any case, there is a direct dependence by forest-dwelling communities on their forest resources.

In contrast to such broad, locally-based needs for forest products, forest departments tend to focus on some highly specialised and marketable forest products, (i.e. on a narrow range of timber species and some few NTFPs to meet regional and national needs).[3] Forests with a homogenous stock of required forest products serve such needs better than mixed forests. The commercialisation of forestry favours homogenous, mature crops to be easily managed and harvested to optimise profits (Subhabrata Palit 1997: 211). The prototype of such market-oriented and highly specialised forest-use systems are plantations. Where plantation replace natural forests used and managed by local communities they may drastically curtail the latter's livelihood.[4] However, over the last two decades there has been a growing awareness in professional forestry about the ecological value of forests. As Ravindranath *et al*. observe: 'Increasingly, foresters are beginning to think in terms of ecosystems rather than managing a few valuable timber species' (1996:306). Efforts to protect forests as ecosystems with a high degree of biodiversity could meet the needs of local people, but in most cases the management of forests for ecological purposes excludes local user groups.

---

3. Poffenberger (1996: 259-86) deals in more detail with the values ascribed to forest resources by rural communities in contrast with forest departments, commercial and other non-local user groups.
4. Robert S. Anderson and Walter Huber (1988) discuss the consequences and conflicts of the replacement of a natural forest with pine plantations for the tribal people in Bastar Central India.

THE INTEGRATION OF FORESTS INTO OTHER NATURAL RESOURCE MANAGEMENT SYSTEMS

A major difference between indigenous and professional foresters refers to the degree of their integration into other natural resource systems. Professional management focuses exclusively on highly specialised working plans based on technical and revenue considerations. Indigenous forest management systems in most cases are closely intertwined with other local management systems of natural resources such as pastures, agricultural lands, swidden plots, gardens, fishing grounds and so on. This requires a different type of decision-making based on a different pattern of constraints set by the requirements of the other management systems. Forest conservation may not be given first priority in the context of scarcity of agricultural land for food production. Open forests allowing good grazing for cattle may be preferred by pastoralists to dense closed forests with good timber. Artisans like potters will be mainly interested in a forest yielding enough fuelwood for their kilns, whereas farmers dependent on irrigated agriculture may be mainly interested in the water retention capacity of forest vegetation. An additional degree of complexity of local systems is due to the fact that even within the same community there may be heterogeneous user groups as in multi-caste villages of artisans, farmers, and cattle herders. In a specific local context we may therefore be confronted with a patchwork of different management systems handled by different social groups, subject to dynamic processes of socio-economic change.

MANIPULATION OF FOREST

Coming to the manipulation of forests as an important aspect of management, we may differentiate the following levels of intervention:

(1) *Low level manipulation*: forest products are appropriated in a controlled manner and there are no conscious attempts to change the natural flora and fauna, its composition and structure.

(2) *Medium level manipulation*: The species composition, structure, density pattern and/or functions of the forest are partially and intentionally changed by human activities.

(3) *High level manipulation*: the composition, structure and/or function of the forest are highly transformed.

However, these three levels of manipulation do not allow any general statements about the quality of the forest. A forest may be enriched in biodiversity by human manipulation or degraded. It may expand in size or be reduced. Manipulation results in change, but the nature and direction of the change is manifold.

Indigenous forest management systems show various degrees of manipulation and controlled extraction of forest resources. Manipulation may be done in the following ways[5]:

1. By protecting selected trees, plants and animals, and/or by reducing competitors, the stock of wild species is manipulated. Such manipulation may be based on material needs to promote the flow of required forest products (e.g. highly appreciated fruit- or sap-yielding trees). But it may also be based on religious beliefs postulating specific relations between certain species of plants and animals with certain categories of people. Totemic meaning of such species may result in taboos and 'social fencing' (Ravindranath *et al.* 1996:297). By controlling the rate of extraction by clearly defined user groups, the quantity and timing of resource extraction can be regulated.

   Silvicultural operations like grafting, thinning, rejuvenating, cleaning, pruning, and improving varieties, used by indigenous forest managers remains a subject of ongoing research. Wiersum points out that. 'Only limited attention has been given to whether and how local people control such utilisation and/or manipulate forests by purposeful practices so as to optimise the benefits from the forests'. (1997:8) The use of fire and other traditional patterns of manipulation are fields for further research. Such research of course includes questions about the criteria used by the local people to manipulate forests in the way they do.

2. Protecting certain forest patches as a whole is another technique found in India to manipulate forests. In such cases, access to certain forest patches is highly restricted as in sacred forests and groves and burial grounds in the forest. However,

5. See also Fernandes *et al.* 1988 (esp. 159-74) and Deeney & Fernandes 1992.

not every sacred grove is untouched, and there are examples of highly manipulated temple-yard forests.[6]

3. The conversion of a forest into agricultural land may be considered a rather high degree of manipulation. Swiddening consists of the clearing of a forest patch in order to grow domesticated plants and crops for a certain period. The techniques of clearing vary. Sometimes selected tree species are not felled at all. The height at which trees are cut also vary between different systems of swiddening which results in different ways of regeneration. The same applies to the period of cultivation, the type of cropping, and the period of fallow. During the fallow period, the natural vegetation recovers and the forest regrows, provided the fallow period is long enough. However, the original vegetation changes its composition, and another type of forest replaces the original one. Domesticated plants may continue to grow in the regrown forest, adding new species to it.
   However, under conditions of scarcity of land and forest resources, shifting cultivation may result in an irreversible conversion of forests into fallow land stocked with bushes and grass.

All three levels of forest manipulation may be found in indigenous forest management systems. In India, tribal communities, who are the main forest dwellers are commonly classified into 5 categories (the number of categories may vary according to the criteria used by the respective author). With the risk of over-simplification, these categories may be linked with the degree of manipulation of forest resources in the following way:

1. *Hunting and gathering tribes*: They are characterised by a high dependence on forest resources for the satisfaction of almost all their needs, but a low degree of manipulation of forests. However, such tribes are a small minority in the Indian context and some of them have developed reciprocal

6. See, the case study of Madhav Gadgil and V.D. Vartak (1998) about sacred groves in the hilly regions of western Maharastra. Field research conducted by the 'Council of Cultural Growth and Cultural Relations' among the Dongaria Kondh and Kuttia Kondh since 1992 in southern Orissa found several cases of sacred groves. A short description of temple forests in India is given by Chandrakanth *et al*., 1990.

relationships with peasant communities with whom they exchange forest products for agricultural produce or other items not available in the forest.

2. *Shifting cultivators*: They are characterised by a high dependence on forest resources for the satisfaction of a broad range of needs coupled with a high degree of manipulation in shifting cultivation areas. Shifting cultivation is still practiced by many tribal groups in India.
3. *Pastoralists*: Their dependence on forest resources may be characterised as 'medium' in the sense that they do not depend exclusively on forests for grazing, but on other lands as well (pastures, wasteland, agricultural land after harvest). The degree of manipulation is between medium and high, depending on the use of fire and the number of animals.
4. *Settled agriculturists*: In comparison to hunter gatherers, shifting cultivators and pastoralists, their dependence on forest resources is less and the degree of manipulation of forests (with the exception of conversion of forest land into permanent agricultural land) is on the medium to low level. Their main dependence is on cattle, the plough, and methods of intensive farming like terrace cultivation and irrigation.
5. Some tribal communities have specialised in basket-making, mat weaving, iron work and so on. They are highly dependent on forest products, but in comparison to hunters and gatherers and shifting cultivators they focus on a narrow range of forest produce. By harvesting a large quantity of the same forest product, they may change forests to some extent. How far they manipulate forests by silvicultural activities in order to favour the growth of the materials required is variable. Probably, their degree of manipulation is at the low and medium level.

Different levels of forest manipulation may exist in combination within a single community in different forest patches. The Dongaria Kondh, for example, practice shifting cultivation and spare sacred groves from any human intervention. In the case of professional forest departments, we will equally find management systems of all three types:

1. A low-level of manipulation is aimed at in the management of national parks, sanctuaries, reserves and other specially protected areas: here the main efforts of the forest departments focus on the control of access to these forests to protect the species composition, structure, flora and fauna.

2. A medium-level of manipulation is to be found in the management of reserved and protected forests managed by the forest departments for various purposes, but especially for timber production.

3. A prominent example of high-level manipulation refers to natural forests replaced by plantations or ones that have been artificially enriched by valuable timber species. Other examples of high levels of manipulation are afforestation projects in highly degraded forest areas.

A systematic comparison between the dominant practices of forest-dwelling communities and forest departments will probably come to the conclusion, that the latter tend to operate more on the medium and high level of manipulation, changing natural forests considerably. In contrast, indigenous forest management systems may be found to operate on the first and second level of manipulation and only in rare cases on the third level, with the exception of communities practicing shifting cultivation.

KNOWLEDGE BASIS

As a rule, local communities manage smaller forest areas than the forest departments. They are also easier to survey and local users often have detailed information about their condition. Local groups dispose of intimate first-hand knowledge about the natural processes of flora and fauna and the local habitat which may go back several generations. Their knowledge of forests is embedded in their whole lifestyle. In this sense it is a rather holistic knowledge.[7]

Large forest areas controlled by forest departments create problems of acquiring detailed information about the condition of the forest resources, as well as difficulties to control access to the

7. However, this does not mean, that everybody knows the same about forests, knowledge about forests may depend on sex, age, occupation, economic situation, etc. There may be specialists and experts within forest-dwelling communities who know more than other common people.

forest and rates of appropriation of its products. Frequent transfers of forest officers render the accumulation of 'local knowledge' difficult. On the other hand, the Forest Department disposes of an expert knowledge based on modern natural sciences. It also has access to information about regional, national and global correlations, (e.g. the functions of forests on a global scale for the climate, or information about non-local demands for certain forest products).

Institutional framework

Indigenous forest management is based on local institutions and operates on locally-accepted norms and values. It is a decentralised management system, adapted to micro-local conditions with a certain degree of flexibility. The user groups and 'managers' are identical. They manage the forest according to their own needs. Decisions, rules and regulations referring to the appropriation and distribution of forest products are made by the local users. Forests used by forest-dwelling communities are, or have been, most commonly used as common property systems, not to be confused with open access regimes. In such traditional common property systems, access to forest resources is regulated by membership in local communities (residence, kinship, caste) controlling specific territories. Arrangements between local communities and non-local user groups (e.g. nomadic groups) may allow the use of forest resources by outsiders not belonging to the community controlling the forest territory. However, there are also instances of 'private property' regimes in forest-dwelling communities. Individuals, households and/or other smaller groups of community members may have exclusive access to, and rights over specific forest resources (hunting game, single trees, beehives).[8]

Forest departments, on the other hand, manage their large forest areas as state property controlling and regulating the access of all forest user groups. This type of management may be characterised as 'custodial management' (Palit 1996:211) with forest officers having policing functions. The forest management of the official departments is based on national institutions, and operates on the basis of government laws, rules and regulations. These restrictions are outcomes of complex political processes on a regional, national and

8. See, e.g., Jena *et al*. (1997:87) for the ownership pattern of caryota urens among the Kuttia Kondh of Orissa.

even international level. Indian forest policy has to try to reconcile many needs and requirements of different segments of a huge population: industrial demands for timber, pulpwood, market demands for construction, fuelwood and NTFP, apart from environmental aspects.

Despite such institutional frameworks in many cases, we have a *de facto* open access situation. Local institutions or common property regimes degrade or collapse due to internal and external pressures due to a growing commercial demand for forest products, population growth, uncontrolled immigration of new user groups, interest groups powerful organised who appropriate forest products illegally. Equally, forest departments in many cases face problems to control forest areas successfully and to implement their rules and regulations.

Joint Forest Management: the merging of two traditions?

Having sketched out the two traditions of forest dwellers and official forest departments, I would now like to discuss some aspects and problems of participatory management approaches. JFM strategies in India are aiming at the regeneration of degraded reserved and protected forests. The official policy of involving local communities in the management of government-owned forest lands is outlined in the 1988 National Forest Policy, and the Joint Forest Management Resolution of June 1st 1990. Both bills recognise that the policy of excluding local people from forest management in order to protect forests has failed. For more than one century of national control over forest resources it had not been possible to exclude people from using the government-owned forests. Due to several reasons, the strategy of social forestry was not able to relieve pressure on natural forests under government control.

JFM strategies, in contrast, bring together local forest users and forest departments, and with them two traditions of forest management. Although there may be similarities in some respect between the two, the there are also differences with respect to needs and priorities; silvicultural techniques and practices; institutional arrangements of access to, and control of forest resources; rates of exploitation of resource units and their distribution; as well as systems of knowledge and know-how. (See *fig*. 1)

According to the degree to which either rural communities or forest departments integrate and realise their own concepts of forest management, approaches involving local communities may be classified into the following three categories, as defined by Johnson (1995):

1. 'Participatory Forest Management (PFM) is where the Government takes initiatives, manages the resource and the community participates in various forms, most commonly as hired labour' (Rural Litigation and Entitlement Kendra 1997:39). Many social forestry projects of the 1970s and 1980s may be classified under this category.
2. 'Joint Forest Management (JFM) is where the owner (the Government) as well as the user (the Communities) manage the resource and share the costs as well as the benefits thereof' (Rural Litigation and Entitlement Kendra 1997:39). This strategy is currently followed by most Indian states (Poffenberger & Singh 1996: 65).
3. 'Community Forest Management (CFM) is the case where the community takes the lead, manages the resource while the government is a passive supporter or observer' (Rural Litigation and Entitlement Kendra 1997:39). This type of management is strongly advocated by environmental and social activists, and NGOs representing the interests of tribal and other forest-dwelling peoples, together with scientists dealing with ecological, social and environmental aspects.[9] However, there are few forestry projects giving that much control and free play to local communities.

In the following discussion, I will focus on JFM strategies as currently taken up by several Indian states and point to some of the main challenges and crucial questions involved in JFM, a strategy which involves the above-sketched traditions of two actors:

1. The first question refers to the compatibility of the needs and management priorities of the two sets of actors. Under current JFM strategies all activities included in the microplans need to confirm to the working plan prescrip-tions, and to

---

9. See, the people-oriented positions represented by various authors in Fernandes (1996) 'Drafting a People's Forest Bill. The Forest Dweller-Social Activist Perspective'.

the Indian Forest and Conservation Acts (Palit 1996:221ff). According to Subhabrata Palit, 'divisional working plans are normally drawn up for a twenty year period and legally dictate what development and management activities are approved in the area' (1996:221).

Working plans tend to emphasise timber production (i.e. species selection and felling cycles (*ibid.*)). The strategy is therefore to embed JFM projects into the general management plan of the forest departments. As long as local needs and priorities do not curtail the policies of the concerned forest departments, conflicts between the two actor groups are avoidable. However, what happens if local communities give priorities to needs which contradict those of forest departments? Are the latter able, for example, to renounce revenue from timber harvest if local communities are not interested in timber species but give preference to non-marketable resources which are of basic value for their livelihood? How far will forest departments make concessions to local forest users who demand improved production and marketing structures for MFP, which will give them necessary cash income but may not yield any monetary profit for the forest department?

Asked the other way round: could it be that working plans will be designed primarily according to the needs and priorities of local forest users? This is what Subhabrata Palit, Chief Conservator for Social Forestry, West Bengal Forest Department, suggests:

> In areas where JFM systems are widely adopted, working plans will become an aggregate of microplans. In such cases they will have little relationship to earlier working plans with their general prescriptions for large areas. Rather, a new analytic and administrative framework for supporting decentralised microplans will need to be created, detailing regeneration methods, development strategies, sharing agreements, and silvicultural operations. Microplans will highlight specific local needs and community development priorities within a broader legal and strategic forest

> management programme, for a region based on community priorities and on changing ecological and economic conditions. — 1996:222

This vision of an Indian forest officer indicates some of the basic structural changes required from forest departments if JFM is to be more than the utilisation of local user groups for ready-made management needs and priorities of the forest departments.

2. A further question relates to the silvicultural techniques and practices used by the two actor groups. There are several challenges linked with the question of which techniques to be followed in JFM. On the one hand, there is a strong demand from NGOs, social and environmental activists and scholars to build on local knowledge and to take local forest users seriously as experts in their own environment. However, in many contexts this approach alone may not be easily implemented or efficient enough for because the techniques and practices of forest dwellers to use and manage their forests are not yet well documented. To make local knowledge the practical and technical foundation of JFM, requires the study of local techniques with regard to their efficiency and sustainability in each area where JFM is to be implemented. To assume that local practices are *per se* the most appropriate solution to a sustainable forest management may prove to be short-sighted in the context of an already degraded forest environment. Local forest users may lack the knowledge or motivation to restore highly degraded forest lands which require more than closing areas to start natural regeneration. Silvicultural operations like the inducement to natural regeneration of existing roots stock; root stock cleaning; thinning of coppice shoots; seeding; gap filling; and intensive planting together with soil and water conservation measures like contour trenching, vegetative bunding, small check-dams, and other technical operations, may be necessary.

   Equally, forest departments with their long-standing experience of timber production lack the knowledge and techniques to manage the types of forests required by local users. Ravindranath *et al.* point to the fact that so far little scientific research has been conducted to identify the most effective ways to manage natural forests intensively for both timber and non-timber forest products. Not much is known about the practical ecological requirements of most NTFPs, their management within a multiple-use forest, yield-data for different NTFPs, or methods to enhance their productivity. Local knowledge about uses of NTFPs, their propagation and ecological requirements could be a valuable starting point in this respect (Ravindranath *et al.* 1996:305ff).

This means that new techniques and practices will have to be developed while not neglecting already existing silvicultural methods, be they traditional methods of local forest users or those of the Forest Department. This requires systematic research and experimentation, and the co-operation of forest departments, local communities, universities, NGOs and other development institutions.

3. My third and last point refers to institutional rules and regulations concerning access to forest resources, as well as the distribution of forest products and benefits. These aspects include questions of legitimacy of JFM in the interest of forest-dwelling communities.

   The basic organisational units involved in JFM are the forest departments represented by the concerned administration units and the Village Forest Protection Committees (in Orissa the Vana Samrakshana Samiti), as well as NGOs. In the Resolution of 1990 of the Ministry of Environment and Forests, GOI, the role and function of NGOs in JFM is considered to motivate and organise village communities for the protection, afforestation and development of degraded forest land. NGOs are expected to act as an interface between the State Forest Departments and the local village communities. Village communities are defined as 'beneficiaries' entitled to share in usufructs subject to the conditions prescribed by the State Government. The ownership of protected forests clearly stays with the State Forest Departments. The Forest Departments are supposed to give financial assistance for necessary silvicultural operations. The role of the Forest Department is to control and supervise the implementation of the Working Scheme, which 'should be prepared in consultation with the beneficiaries'. In case the beneficiaries fail to perform the operations prescribed by the Working Scheme, or to protect the area efficiently, forest departments may withdraw the entitlement to usufructuary benefits without paying any compensation.

The constitution of Village Forest Protection Committees raises several issues challenging not only forest departments but local forest users as well. Such issues relate to the criteria of membership to join forest protection committees; the identification of the forest area to be managed under JFM; and the negotiation among local forest users about the needs and management priorities covered under JFM. Although in some cases there may be well-established and organised local user groups based on traditional structures with convergent needs and interests as well as clearly defined rights to use certain forest patches; in many cases the situation will be characterised by one or more of the following problems.

Due to the prevailing open access situation in government forests, traditional common property regimes have collapsed. Forests are used by a wide range of heterogeneous user groups, by locals as well as outsiders. Headloading of fuelwood and collecting *sal*-leaves, may be done by villagers not residing close to forests and nomadic pastoralists who only use local forests periodically. By giving exclusive usufruct rights to Village Forest Protection Committees, such users will bear the costs without getting any benefits from the in protected forests. Neighbouring village communities having used the same forests for long periods may dispute the claim of one village to manage a particular forest under JFM.

Due to processes of socio-economic differentiation, needs and priorities for forest resources within a specific local community may be in conflict, and it may not be possible to come to an agreement regarding the communal management of a patch of forest. In heterogeneous multi-caste villages such diversity may also impede the communal management of a forest under JFM as the more powerful segments of the village community may dominate the weaker sections, which are highly dependent on forest resources.

Forest departments may therefore be confronted with various demands for the establishment of a community forest. Experience in eastern India has shown that spontaneous community forest initiatives have been based on clustered hamlet groups of 20 to 50 households, most commonly belonging to the same tribe, clan or caste (Sarin 1996:172). Such small groups based on traditional social structures turned out to be better suited for JFM than large village forest protection committees initiated by NGOs and forest departments. However, it is doubtful whether the administrative expenditure for the Forest Department to deal with hundreds and thousands of small forest protection committees, to elaborate micro-plans and to supervise and monitor their implementation is a viable solution under existing administrative structures. The question would therefore be, how the forest departments are to be re-organised to fulfil such duties, or who else, and what type of institution could resume such duties.

The benefits accruing from the protection and regeneration of degraded forests are a further critical aspect of JFM strategies. If, through JFM, local user groups are legally entitled to what they used to take from protected and reserved forests, and they have to expend extra time and energy in regeneration activities, the incentive of legalisation may not be enough to stimulate JFM. Thus successful JFM agreements between forest departments and local user groups require a long-term security to usufruct and benefits for the Forest Protection Committees. Last but not least, the size of the forest area to be managed under JFM should be large enough to fulfil the needs of the local user groups, otherwise pressure on forest resources will be shifted to other forests which are not under JFM.

Such a forest policy, however, aim to support and develop forest-dwelling communities, giving them exclusive usufruct rights to use forest land and resources. In doing so, the government may well be confronted with the claims of other interest groups such as industrialists, urban consumers, or other rural communities. If JFM strategies are considered to be 'people-oriented' approaches, it must not be forgotten that forest dwellers are only one part of Indian society. How legitimate is the preferential treatment of forest-dwelling communities?

As I said in the beginning, I do not have answers to all these questions. JFM is not the merging of two traditions of forest management, but a new management system requiring new skills and capacities from the forest departments, as well as from the forest-dwelling communities. However, there is a chance of mutual learning. The evolution of JFM as another forest management tradition will be a long process, involving progress and setbacks, conflicts and solutions. In the context of shifting power-relations, the forest departments may lose their long-established policing power, but they may gain the co-operation of local forest dwellers to find new solutions to the old problem of how to manage the forests in a sustainable way. JFM strategies which try to impose the management concepts, of forest departments on forest-dwelling communities will fail.

In my opinion, a close co-operation and dialogue between village communities, Forest Departments, NGOs, scientists and development institutions is necessary to move towards the creation of sustainable management systems which do not exclude those who are depend immediately on forests for their livelihood. JFM strategies are the first step in this direction. However, there are still many problems to be solved:

- Scientists from various disciplines (e.g. botany, ecology, anthropology and sociology, forestry, economics and political sciences) are required to study the various aspects of existing forest management systems to provide the necessary data by which solutions may be outlined.
- If NGOs will be further involved in strategies of JFM, they will have to be able to deal with a broad range of conflict-resolution strategies to mediate between different forest user groups, as well as between the Forest Department and the user groups.
- Of all the parties involved in JFM, the forest departments will be confronted with the biggest challenges. If taken

seriously, JFM requires an almost completely new attitude towards local users, requiring changes in their management approaches as well as administrative reorganisation.

- Local user groups are equally challenged to change their ways of using forests. Traditional techniques, practices and institutions may no longer be efficient in the context of degrading forest environments. Local users will have to develop new skills and capacities to deal with the forest departments, and they will have to bear the opportunity costs for protecting and regenerating forests.
- Local, national and international development institutions will have to support this complex process of evolving new management systems and act as an interface between the different actors and their interests.

The dialogue about forest is a dialogue between many actors having different ideas and perceptions of trees and forests, with various practices, needs, values and knowledge. But all talk about trees and forests. To listen and to learn about the other's 'tree' and 'forest' is a basic requirement for a dialogue and practice which aims at the preservation of forests not only as providers of valuable resources, as living space for flora and fauna, as most precious ecosystems, but also as a living space for the local forest dwellers.

Fig. 1: Comparison of two management traditions and changes required by JFM

Aspects of forest management system	*Forest-dwelling communities*	*Forest departments*	*Changes required by JFM*
Needs	**May vary from multiple to specialised material needs**: fuel, fodder, manure, medicine, building material, grazing area, hunting area, agricultural tools, handicraft	**Predominantly specialised material needs:** marketable timber species and some few MFPs	FDs are required to reorient management emphasis to multiple-use, non-timber forest production
	Socio-cultural needs: hunting ground, burial place, sacred groves, forest as living space	**Ecological needs:** protection and conservation of forests as ecosystems	Where FDs protect and conserve forests as ecosystems, they will have to include local users.
Degree of integration into other natural resource management systems	**From low to high:** *low*: hunters and gatherers *medium to high*: shifting cultivators, settled agriculturists, pastoralists	**Low**	FDs are required to take into consideration other natural resource management systems. In highly degraded areas, local user groups may have to give more emphasis on forest management and renounce short-term benefits from other natural resource management systems
Degree of manipulation	**From low to high:** *low*: hunters and gatherers *medium*: shifting cultivators, pastoralists, settled agriculturists	**From low to high:** *low*: national parks, sanctuaries, reserves and other specially protected	The degree of manipulation required in JFM may vary from low to high: badly degraded areas

	high (under certain conditions): shifting cultivators, pastoralists, settled agriculturists	areas *medium*: reserved and protected forests *high*: plantations, reforestation of highly degraded forest areas	may need a high degree of manipulation not familiar to local users, whereas the regeneration of less degraded natural forest may be done with a low or medium degree of manipulation.
Knowledge	**Local and intimate knowledge embedded in lifestyle, holistic**	**Generalised knowledge together with highly specialised knowledge**	JFM strategies require a combination of local knowledge and the knowledge of FDs. However, further research is necessary to create new knowledge and know-how, especially regarding the sustainable management of forests for multiple use and NTFP production
Institutional Framework	**Common property systems**: based on locally-accepted norms and values, access to forests via membership in local communities (residence, kinship, caste) decentralised, flexible	**State property system:** custodial management based on government laws, rules and regulations, bureaucratic, centralised	A new institutional framework has to be created under JFM, based on existing local structures. The problem of legitimacy for the FDs has to be solved with local user groups, JFM to be backed by official legislation (Forest Act, Forest Policy, laws, rules and regulations)

References

Anderson, Robert S. & Huber, Walter, 1988, *The Hour of the Fox. Tropical Forests, the World Bank and Indigenous People in Central India,* Seattle and London.

Chandrakanth M.G., I.K. Gilless, V. Gowramma, M.G. Nagaraja, 1990, 'Temple forests in India's Forest Development', In: *Agroforestry Systems, 11*, pp. 199-211.

Colombijn, Freek, *Sustainable use or reap-and-run, views of the Indonesian forest by locals, outsider and the state, 1800-2000.* Paper presented at the First ICAS Conference 'Changing Asian Forests', Noordwijkerhout, 25-28 June 1998, pp. 1-20.

Deeney, John & Walter Fernandes, 'Tribals: their Dependence on Forests, their Traditions and Management Systems', In: Fernandes, Walter (ed.), 1992, *National Development and Tribal Deprivation,* New Delhi, pp. 49-75.

Duerr, W.A. *et al.* (eds.), 1979, *Forest Resource Management, Decision-Making Principles and Cases,* Philadelphia.

Fernandes, Walter, Geeta Menon, Philip Viegas, 1988, *Forests, Environment and Tribal Economy. Deforestation, Impoverishment and Marginalisation in Orissa,* Indian Social Institute, Tribes of India Series, New Delhi.

Fernandes, Walter, (ed.), 1996, *Drafting a People's Forest Bill. The Forest Dweller-Social Activist Alternative*, New Delhi.

Gadgil, Madhav & Ramachandra Guha, 1996 (1992), *This Fissured Land. An Ecological History of India,* Delhi.

Gadgil, Madhav & Ramachandra Guha, 1995, *Ecology, the use and the abuse of nature in contemporary India,* New Delhi.

Gadgil, Madhav and V.D. Vartak, 'The Sacred Uses of Nature', In: Ramachandra Guha (ed.), 1998, *Social Ecology,* Delhi, pp. 82-89.

Government of India, Ministry of Environment and Forests, Department of Environment, Forests and Wildlife, Resolution, No. 3-1/86 FP, 7th December, *National Forest Policy, 1988.*

Government of India, Ministry of Environment and Forests, Department of Environment, Forests and Wildlife, Letter No. 6-21/89-F.P., 1st June 1990, Subject — *Involvement of Village Communities and Voluntary Agencies for Regeneration of Degraded Forest Lands.*

Government of Orissa, Forest, Fisheries & A.H. Department, Resolution, No. 10 F (Pron) - 47/88 /17240/ FFAH, 1st August 1988, Subject — *Protection of Reserve Forest Areas by the Community.*

Government of Orissa, Forest, Fisheries & A.H. Department, Resolution, No. 10 F (Pron) - 47/88/23638/ FFAH, 13th October 1988, Subject — *Protection of Reserve Forest Areas by the Community.*

Government of Orissa, Forest, Fisheries & A.H. Department, Resolution, No. 10 F (Pron)-4/90/29525/FFAH., 11th December 1990, Subject — *Protection of Reserved Forest and Protected Forest Areas by the Community and Enjoyment of Certain Usufructs by the Community.*

Government of Orissa, Forest and Environment Department, Resolution, No. 16700—10-F (Pron)-20/93-F. & E, 3rd July 1993, Subject — *Protection of Reserved and Protected Forest Areas by the Community and Enjoyment of Certain Usufructs by the Community.*

Government of Orissa, Forest and Environment Department, No. 14292—10 F (Pron) 1/94-F. & E. Corrigendum, 4th August 1994.

Jena, M.K., K.K. Patnaik, K. Seeland, 'Where trees do matter for society: the cultural aspects of sal (*Shorea robusta*) and salap (*Caryota urens* L.) in the Similipal hills of Orissa, India', In: Seeland (ed.), 1997, *Nature is Culture. Indigenous knowledge and socio-cultural aspects of trees and forests in non-European cultures*. London, pp. 79-89.

Johnson, S., 'Clarified JFM Concepts', In S.B. Roy (ed.), 1995, *Enabling Environment for Joint Forest Management,* New Delhi.

Palit Subhabrata, 'Indian Forest Departments in Transition', In: Poffenberger & McGean, 1996, pp. 210-29.

Poffenberger, Mark & Betsy McGean, (eds.), 1996, *Village Voices, Forest Choices. Joint Forest Management in India,* Delhi.

Poffenberger, Mark, 'Valuing the Forests', In: Poffenberger & McGean, 1996, pp. 259-87.

Ravindranath, N.H., Madhav Gadgil, Jeff Campbell, 'Ecological Stabilization and Community Needs. Managing India's Forest by Objective', In Poffenberger & McGean, 1996, pp. 287-323.

Sarin Madhu, 'From Conflict to Collaboration: Institutional Issues in Community Management', In Poffenberger & McGean, 1996, pp. 165-209.

Rural Litigation and Entitlement Kendra (ed.), 1997, *Community Forest Management in Protected Areas*. Van Gujjars Proposal for the Rajaji Area. Natraj Publishers, Dehra Dun.

Wiersum, K.F., 'Indigenous Exploitation and Management of Tropical Forest Resources: an Evolutionary Continuum in Forest-People Interactions', In: *Agriculture, Ecosystems & Environment*, No. 63, 1997, pp. 1-16.

9

What Alternative Would Conserve Forests Better?

Empirical Studies on Joint Forest Management in Orissa

Padmini Pathi

Introduction

JOINT Forest Management (JFM) is an initiative taken up by the State Forest Department to involve the forest-dwelling communities with the broader interest of conserving forests sustainably. It seems to be an integration of people's indigenous system of forest management and the policy management adapted by the government. At the moment when forest conservation has drawn worldwide attention, JFM has been recommended by the Union Government Ministry of Environment and Forests to the State Forest Services as a new forest policy. When the task of protecting forests became so urgent, the State Governments were increasingly prepared to involve the rural and tribal communities who reside in and around the forests, and favour a more people-oriented management.

JFM has been implemented in various parts of tribal Orissa. Through the formation of Village Forest Protection Committees (VFPCs) and *Vana Samrakshana Samiti* (VSS) — both terms may be brought under the broad term Forest Protection Committees or FPCs — local people have been motivated to manage their forests, on their own, in accordance with an officially-approved forest management plan. Under this policy, local people have been privileged with certain rights and concessions with regard to the management and utilisation of the forests allotted to them. Communities which participate in the smooth implementation of JFM through micro-planning exercises for local economic development are enjoying certain entitlements to usufruct forest produce to satisfy their needs.

Although the Orissa Forest Department has tried to spread JFM all over the State, the concept has not been able to convince and motivate all forest-dwelling communities to co-operate. Positive responses can be seen in north-eastern and eastern Orissa. In these areas, people have become aware of the benefits they are allowed to draw from forest protection and management. In southern Orissa, JFM is a completely new thing for the tribal communities.

In order to understand the different response to JFM by the communities, the author conducted field studies in several areas of northern Orissa. Sampling of the villages was done in consultation with the Divisional Forest Officers (DFOs) of the respective forest divisions. Many micro plans have been submitted to the concerned DFOs by the villagers and communities who are willing to take up forest protection measures. The micro plans are prepared on the basis of the principles of JFM. Till now, about 1105 FPCs have been established in the State who have taken the responsibility of protecting and rejuvenating by enrichment planting of 104 454 hectares of forest land. However, the DFOs and Forest Range Officers often suggest particular village localities where a community has shown interest and response to forest protection. On the basis of such information, fieldwork was done in Dhenkanal, Kendujhar and Mayurbhanj Districts of Orissa. Through direct participation with local communities, group discussions, and studying facts from village proceeding registers, data were collected on communities' responses, and attitudes towards JFM, and ideas of alternative measures. The scope of the paper covers six case studies of FPCs in northern Orissa.

Forest policy background

Scientific forest management based on official forest policies was begun a century ago. Until the 1988 Forest Policy, the official Union Forest Policy, based on the Forest Act of 1927, used 23 per cent of India's forests for commercial purposes. Local communities residing in and around forests were at least given concessions to use and manage forests for their own requirements, while the government was earning substantial revenues from forests and forest produce. Large-scale commercial interests of the government ignored the living conditions of the local people and their immediate needs for subsistence. The tribal forest-based communities and the non-tribal scheduled castes who rely on the marketing of forest produce were confronted with increasing poverty linked to the degradation of forests. Many of tribals, whose socio-economic and livelihood system is the forest, put more pressure on the already fragile forest ecosystem leading to further degradation. The struggle of poor people for survival on the one hand, and the preservation of the environment on the other, became increasingly important issues.

A time came when environmentalists realised that the old strict custodial policies of forest management were counter-productive and policies needed to be changed for the sake of the people, forest and environment. In the 1970s, the achievements of the Arabari pilot project in West Bengal revealed the importance of involving village communities in the protection of natural forests. Subsequently, the CHIPKO movement in Garhwal Himalaya illustrated how local communities were able to the maintenance and regeneration of their natural forests. The new forest policy introduced in 1988 shifted its focus from 'commerce' and 'investment' to ecology and satisfying the minimum needs of the people, and strengthening the tribal-forest linkages.

In June 1990, the Government of India issued guidelines to all States to adopt JFM, measures in which people would protect forest lands and help in their natural regeneration in exchange for rights to collect dead wood and non-wood forest produce, and a 25 per cent share from the marketing of timber. Prior to 1864, tribal community, forest produce management was practically autonomous in many Princely States and in British-India. At this time, tribals only had to seek permission to use forest produce and pay certain royalties. With the first Indian Forest Act of 1865, this community management was abolished and replaced by a state forest administration. The increasing scarcity of forest produce severely affected the forest-dependent communities. One hundred and thirty years later the depletion of renewable natural forest resources has led to crises and called for community involvement in the protection of forests to ensure the availability and sustainability of renewable natural resources. At a national workshop on JFM at Surajkund in 1992, for example, the positive and negative aspects of community participation were discussed.

Community participation : Positive aspects

1. Most effective local institutions develop in small communities where people know each other and collaboration is easier.
2. In upland villages common land is visible from most of the dwellings so that unauthorised felling can be easily controlled.
3. Hills and uplands have a larger variety of resources and forests which is vital for local people's survival. This encourages them to engage in protection and management.
4. Remoteness and great distance from roads and markets helps in maintaining mutual obligations and discourages poaching.
5. In remote villages fear of reprisals from village elders deters frequent abuse of common resources.
6. Upland settlements are more homogeneous in caste, with one caste or tribe dominating both in land ownership and number, thus leading to effective social control.

Community participation : Negative aspects

1. The village settlement area in some regions may spread over a length of 5 to 10 km which makes control difficult, and allows illegal resource users to escape undetected.
2. Once degradation in an area starts, the community becomes indifferent to protection.
3. The old system of authority in the modernised villages has been undermined, resulting in a loss of confidence.
4. In multi-caste villages social control is more difficult.

Community control is more likely to work where the organisational level is high, both among the village elite and the commoners. Where the Forest Department and communities are effectively working together to restore the productivity of forest lands, potential returns from such efforts are remarkable. In comparison to the earlier custodian forest policies, and in due contrast to the notion that forests are a state-owned property, JFM policy is people-oriented. The resentment which the forest dwellers felt against the Forest Department's control over forests for so long is now ceasing, at least in areas were JFM is working well i.e. contributing to the livelihood of marginal people). This represents a fundamental shift in forest regeneration methods:

From	To
1. Centralised management	1. Decentralised management
2. Revenue orientation	2. Resource orientation
3. Production motives	3. Sustainability
4. Single products	4. Multiple products
5. Large working plans	5. Micro-planning
6. Target orientation	6. Process orientation
7. Unilateral decision-making	7. Participatory decision-making
8. Punitive roles	8. Self-abnegation rules
9. Controlling people	9. Facilitating people
10. Departmental	10. Peoples' institutions
11. Assumed homogeneity	11. Recognised diversity
12. Achieving single, pre-set objectives	12. Fulfilling multiple need-based objectives
13. Area management	13. Site-specific management
14. Timber production	14. Multiple products combined with bio-diversity
15. Single technical package	15. Menus of options
16. Fixed procedures	16. Experimentation and flexibility
17. Plantation as first option	17. Low-input management and regeneration
18. Single species	18. Multispecies and multitier plantations

(Society for Promotion of Wastelands Development (ed.) 1992:10)

Implementation of JFM policy

Due to the diversity in cultures and traditions in Orissa, traditional modes of management of forests in their socio-cultural contexts are also diverse. Through community participation there is a possibility of integrating these traditional systems of management into JFM. The mode of implementation should therefore not be universal or homogeneous. The *modus operandi* of the management system, wherever possible has to be focused on the social and natural conditions at the local level. The implementation procedure has to be responsive to local needs, culture, tradition and institutions. However, before implementing policy programmes, one must be aware that JFM is working more smoothly, where the social and caste composition of the people is homogeneous, and where they are still substantially dependent on forests. Where villagers have recognised their resource supply crisis and are willing to protect the remnant forests, ecological conditions like the vegetation type, the availability of root stocks to facilitate quick regeneration, and a reasonable land-to-man-ratio should be considered. Interaction between the community, the Forest Department and the local NGOs involved as mediators, should ensure a better protection and management of forests. Local institutions are indispensable to help in the protection and management of forests. The local community's religious feelings towards forests and their supernatural aspects are essential in maintaining forests in general, and sacred groves in particular. These mores and values are indirectly encouraged in the policy issues of JFM.

Case studies

The following case studies (Table 1) discuss the implementation of JFM policy at the local level. The prerequisites of establishing FPCs; their organisational structure adhering to policy issues; and the changes in the local, political, cultural and economic factors influencing the management and protection of the forests are all discussed.

MAHAPADA FPC

Mahapada is a small village located approximately 4 km from Joranda (Dhenkanal District). It has 22 households out of which 20 households belong to the kshatriya caste and two households are brahmanas. The forest which the Mahapada villagers are protecting is denominated as 'Rupabalia B class Reserve Forest - North' measuring about 75 acres. The 1,000, hectare Rupabalia Reserve Forest is located on a hilly ridge surrounded by 11 villages. Villages in the area also protect 700 hectares of undemarcated protected forests (khesra) neighbouring Rupabalia Reserve Forest (Poffenberger *et al*.: 53). The area was originally settled by people of the Saora, Munda and Santal tribes in the late nineteenth century. Later on, non-tribal settlements were established and forest openings in the plains were cleared for agriculture. Brahmanas began moving into the area, often bringing with them families from agricultural and artisan castes, and tribal land rights were neglected in the process. While most tribals have turned to sedentary agricultural life, often as landless labourers, they remain much more dependent on the forests for timber and non-timber forest products than their Hindu neighbours (*ibid*.: 42).

The FPC in Mahapada was constituted on 20 March 1992 when a map of the specific area assigned to the FPC for protection was handed over to the committee by Forest Department officials. The villagers became aware of the opportunity to establish a FPC in 1989 through the Forest Department and one of the villagers, Raghunath Hota, who was then an Additional Conservator of Forests. The Executive Committee (EC) consisted of the local *sarpanch* as the President, the Revenue Inspector, the VLW and seven villagers, none of them were women. The committee decided to hold regular meetings once in a month to plan actions and prepare strategies to protect the forests. The *sarpanch*, Revenue Inspector and the VLW did not regularly attend the meetings, and owing to the feet that the *sarpanch*, had become a member of the EC, issues became politicised.

It was the opinion of the villagers of Mahapada that because the *sarpanch* was a people's representative and a member of a political party at the same time, he could never be a neutral person. It was stated that he had an interest in votes, first and foremost. People felt that, he would try to be fair with all villagers, but as the president of many FPCs under the same G.P. (Gram Panchayat = village government) his decisions could never favour all FPCs of the same locality. When the villagers were trying to exclude the *sarpanch* from the VFC, the State Government put forward the VSS policy. As per the guidelines of the VSS, a VSS executive committee was constituted in the village on 12 March 1996. In the wake of this new policy, the *sarpanch*, Revenue Inspector and VLW were excluded from the EC, and the forester and forest guard became members of the EC who did not have voting rights. The present VSS EC in the village has 12 members including a forester, forest guard and two women. The general body consists of 44 members with one male and one female member from each household.

Working principles

To protect the forest, patrolling is done by two persons on a rotational basis, and during the agricultural activities patrolling is done by a wage labourer who is paid Rs.15 per day. The payment is done from the village fund. Every household contributes Rs.5 per month to the village fund. The EC keeps records of each meeting and from time-to-time they change the plan of management and protection as desired. They have laid down rules that if a person on his duty day did not patrol he will then have to do patrolling for two days as compensation. Every villager is expected to help the patrolling guard and act on his information (in case of smuggling, cutting of wood, etc.), and a fine of Rs. 5 is charged if someone does not co-operate. The patrolling guard is held responsible for damages to the forest during his duty. Persons who collect forest produce must deposit a reasonable amount of money with the EC, and without the permission of the EC one cannot collect forest produce. If someone does so without prior permission of the EC, he will be levied with a fine of Rs.15 and all his collections will be seized.

At regular intervals dry branches and leaves are collected from the forests and shared equally among the households. Any person requiring an additional share has to pay double the amount of one share (fixed by the EC) to the EC. Further entitlements for members of the FPC to forest products are:

- 10 headloads of fuelwood per family per month. On special occasions like a marriage festival, a family is permitted to fetch fuelwood as much as it requires.
- In case a house is ruined by a natural disaster, a family is provided with construction materials from the forest.
- No family can sell any forest produce (even from his share) to anybody inside or outside the village.
- The EC demarcates certain patches in the forest for collective exploitation once or twice a year within a stipulated period, failing which one loses one's chance.
- Attending VSS meetings is obligatory and Rs.5 is charged as a fine if someone opts out.
- If anybody withdraws his membership from the FPC he will not be entertained further as a member.
- If anybody collects construction materials without seeking the permission of the EC of the FPC, he will be charged

Rs.30 as a fine and all his collections will be seized by the EC.

- Nobody can purchase logs and poles from other villages, but people from other villages can purchase forest produces from the EC of FPC of Mahapada.
- The VSS members can fetch 50 small poles and 10 bundles of sticks per year.
- Sweeping the forest floor for dry leaves is done twice a week on Thursday and Sunday with the close supervision of the EC of FPC and the collection is equally shared.
- A person requiring extra materials will have to pay the following rates:

Material	*Quantity*	*Charges (Rs.)*	
		Till 1995-96	*After 12 March 96*
Fuelwood	One bundle	0.60	1.00
Sticks	- do -	0.50	1.00
Small poles (I)	1 No.	0.10	0.50
Small poles (II)	1 No.	0.50	1.00
House-building poles	1 No.	2.50	5.00
Loft-making pole	1 No.	0.50	2.00
Tarva (Pole)	1 No.	0.50	1.00
Semi pole	1 No.	1.00	5.00

According to the older generation in Mahapada, the surrounding forest was very dense during the king of Dhenkanal's period. The king then appointed forest guards to check the forest exploitation by the rural communities, end local people were forced to pay taxes for using forest products. According to Poffenberger *et al* :

> The Rupabalia Forest was originally controlled by the Raja of Dhenkanal who had forest guards living in Nathua and Vejibula villages.
>
> [...] ... after Rupabalia hill was designated a reserve forest, community control broke down

> clear felling of Sal and mixed forests from hills and plains deteriorated the forest condition. By early 1970, most of Rupabalia hill, including the lower Sal forests and upper-mixed forests, were barren. (1996:56)

Prior to the initiation of protection as Reserve Forests (1972-79) communities gradually began to extend control over forests starting in the 1960s. Nathua and Mahapada were among the first villages to control access to the areas designated as Reserve Forests more than a decade later (*ibid*, p. 57). During that time, people from many villages such as Mahapada, Naliragada, Tangarmundi, Harijan Sahi, Vejibula were dependent upon the same patch for all kinds of needs, and therefore protecting the forests. But solidarity among the villages could not be maintained as various villages exhibited different forms of dependence on the forests. Exploitation of forest resources varied in relation to their economic condition and availability of land resources. Finally, the villagers decided to divide this patch and distribute the fragments to the respective villages. Mahapada was the first village to constitute its own FPC and manage its part of the forest on its own.

CHHATIA FPC

Chhatia is a big village comprising of 85 households from various castes like brahmana, kshatriya, Kharia (S.C.), Sabar (S.T.), and Juang (S.T.). The revenue village is divided into four hamlets. The first hamlet called Baula Sahi or Paika Sahi has 14 households; the second hamlet (Majhi Sahi) has 12 households; the third hamlet (Upara Sahi) 30 households; and the fourth hamlet (Tala Sahi) comprises 29 households.

According to the local people, they have been protecting the forest around their village since 1960, but for various reasons the forest deteriorated gradually. Roughly two decades ago the villagers took a bigger interest in protecting the forests with better community involvement. Most of the families in the village were still earning their livelihood from the forests, their main source of income being *kendu* (*Diospyros melanoxylon*) leaf collection, and fuelwood collection along with other MFP available in the forest. This kind of exclusive dependence by certain families on the forests is partially responsible for their deterioration.

Mr. Maharaj, a forester of Deogaon forest office, motivated the villagers to establish a FPC in 1989. Further meetings organised by the Forest Department in the same year enlightened the people about the need of protecting forests and mobilised them to partcipate in conservation measures. In the same year, by oral agreement with the forest officers, the villagers was allowed to protect 2200 acres of Rupabalia B class Reserve Forest including 100 acres of Khesra forest. Thus the FPC was constituted. An influential person in the village, Raghunath Biswal, headed the Executive Committee. Biswal was the supreme and all powerful leader in the EC and it seems as if other members had no power of speech or action.

After 3-4 years the FPC was dissolved due to collective allegations against Biswal. The villagers became dissatisfied with him, because he misappropriated funds and abused his power. Many times Biswal released wood cutters, who the villagers nabbed, without any penalty or punishment. Then N.C. Puhan took over from Biswal, but in due course of time Biswal became successful in dissolving the FPC. In 1995, the FPC, then turned into VSS, was reorganised under the presidentship of Biswal. The Executive Committee was constituted with 12 members including two women. But in the general body only one adult male from each household became member irrespective of the VSS rules to include one adult male and one adult female from each household. Although the FPC was run according to the guidelines of the government, there were many intra-village conflicts regarding the sharing of forest produce, the collection of firewood and construction materials. Conflicts led to a break down of the FPC (VSS) and the assigned forest was divided into 5 parts. The village was also divided into 5 parts, made up of 4 hamlets and the village centre.

This division was necessary because different castes and tribes have different practices to use and manage forests. The Sabar and Juang subsist mainly on wood cutting and selling. Scheduled caste communities depend on wood selling and cultivation. The brahmana and kshatriya families (constituting the village centre) depend on forests for fuelwood. The tribals, in turn, collect fuelwood for those castes. One interesting incident happened when the researcher was requested by Mr. Biswal to ask some women headloaders why they cut wood. The women answered that they had collected it for Biswal's family alone. This exemplifies how members of the high castes on the one hand employ the tribals to cut fuelwood for them, and on the other hand accuse them of destroying the forest.

With the division of the village, and the population of each hamlet being supposed to protect its respective forest, each hamlet has appointed two patrolling guards. The operational principles of the VSS emphasise that the EC of FPC shall act as the decision-making authority. It allows the families to collect fuelwood to meet the household consumption needs, one cart load of wood per family per year and extra loads against a fixed amount of money deposited with the FPC. Penalties of Rs.100 to Rs.200 were to be levied on wood cutters who did not have a permission from the EC. Smugglers were penalised, their wood loads seized, and handed over to the forester. The patrolling guard would be deprived of collecting fuelwood from forest for two days, if he neglected his duty.

In the protection of forests and patrolling during the day-time tribal women are much involved. Tribal men do the patrolling at night and the patch of forest land which is given to the tribals is well protected by them. Quite they have often no idea of what VSS and JFM is all about, but they believe themselves to be the traditional conservators of the forest.

Katalposi FPC

The village of Katalposi is under the Jhumpura block in Kendujhar District. The village has 50 households with a population of 298 people. Out of 50 households, 39 belong to the Mahanta (Kudumi) caste, and 11 are Majhi (Santal, a scheduled tribe) households. The forest which they are protecting is classified as 'Balibandha 'B' class Reserve Forest' covering an area of 50 hectares including 15 hectares of Khesra forest.

In the early 1960s the villagers became conscious about the need to protect and manage the forests they use, because they were facing shortages in forest products. They looked after the forests in their traditional way till 1986 when they heard of FPCs. The people felt encouraged to establish a FPC under the supervision of the Forest Department which motivated people with great care to participate. In the beginning, there was a conflict between the two parties of the same village who claimed their customary right to use a particular patch of forest. Further conflicts challenged the position of the president in the FPC. However, with the intervention of the Forest Department the first FPC was constituted in 1987 with 9 members (a convenor; a secretary to keep the records; a general secretary to deal with the administration as per the government guidelines; a liaison man between the Forest Department and the village; a male and a female member of the Majhi community; as well as 2 male members and one woman from the Mahanta community).

The main objective of the FPC was patrolling. On a rate duty basis two persons from two different families were to do the patrolling. Only one person from each family was allowed to collect leaf litter and fuelwood. Mid-term felling at an interval of 5 to 6 years and equal sharing of timber among all families in the village was agreed upon. This was in contrast, however, with the people's traditional rights to exploit Khesra forests.

With an increase in population and the splitting up of families due to marriage and the setting up of separate households, the number of individual shares increased, yet the share of forest products decreased. To overcome this shortage, the villagers became even more serious about the protection of the forests, and with regular patrolling they stopped the illegal felling of trees and the exploitation of their forest by the neighbouring villagers of Bandha Sahi and Dhepa Sahi.

The people in Katalposi were paying nistarcess (royalties). The Forest Department gave a utilisation permit to exploit forest produce from the Khesra forest. The people thus used to pay one quarter of the estimated cost of their collection for which the exploitation was less limited. Now, with the constitution of the FPC, accessibility to Khesra forests is also prohibited.

In 1993, the villagers constituted VSS as per the guidelines of the Forest Department. The EC was constituted with 12 members who were voted on directly by the people in the village (the Naib Sarpanch as the Chairperson, Ward member, three women, five men, forester and forest guard). However, no voting was required for the forester and the forest guard who were included in the EC. From every family an adult male and female became members. The people in this way realised a change in organisational structure, involvement and benefit sharing between the previous FPC and the VSS. The major responsibility of the new VSS became patrolling, silvicultural practice, penalty collections, reporting offences to the Forest Department, overall supervision of protection, and the equal distribution of forest collections among households.

For a better management the VSS now imposes restrictions on people to commercially exploit brush sticks. Emergency meetings can be convened at any time in which all village members must participate. Emergency meetings are generally organised when wood smugglers are detected. No person can collect anything from the forest without permission of the EC. In case of a forest fire all villagers must try to extinguish it, and to prevent further fires, the forest floor must be cleaned everyday. One cannot use hired manpower for MFP collection.

The villagers have appointed two permanent guards for patrolling. Everyday one of them must accompany the two village patrolling guards. As remuneration for the guards, every household must contribute 20 kg of paddy per year.

The people, however, in a self-evaluation of their management feel that the operation of the FPC is very far from conserving the forest for future use. Yet, the forest at present is in a better condition as compared to that in the past.

JAMUNALIA FPC

The village Jamunalia is of great interest in the context of JFM, for there is a 'women management' system in operation. The village comes under Saharapada block of Kendujhar District. It can be approached by following the route running from Dhenkikole of Kendujhar to Mayurbhanj. The total number of households in the village is 152 including 25 Munda families, 100 Gond families, 25 Mahakudas (milkmen) and 2 families of blacksmiths. There is a FPC (VSS) in the village which enrolls 107 households. Fifty-five households have been generated out of family fission and will be included in the list after paying their membership contribution to the village fund.

The idea of forest protection in Jamunalia village appears to be very traditional as well as based on the realisation of a crisis when the forest area fastly degraded due to overexploitation. To ensure the supply of forest produce to the household needs in perpetuity, the villagers made all possible efforts to conserve and protect the remnant forests. The villagers report that three decades ago there was deep dense forests around them and several villages (viz. Jamunalia, Tangarapada, Javasura, Badbil and Dasapada) were depending on this patch of forest. Mining activities and industrialisation in Joda and Badbil led to the deterioration of these forests and the villagers of Dasapada and Tangarapada began to exploit the forest commercially. This situation made the villagers of Jamunalia encroach on a big patch of forest in order to protect it.

This encroachment of forest land and its active protection and management was the beginning of a VFC formation. In 1984 the people constituted an informal VFC at the village level with active co-operation from the Forest Department and its officers who took an appreciable interest in motivating and mobilising the local people. The villagers of Jamunalia were accorded to protect 38 hectares of forest around their village known as Budhakamana (10 ha), Chhanakamana (25 ha) and Masanipada (3 ha). The Masanipada is a sacred forest where the community worships and performs festivals. On normal days nobody enters into the Masanipada forest. The above-mentioned forests are not reserve forests, or parts of them, but are regarded as village forests.

After constituting a VFC in the village, the people became more responsible about the protection and management of the forests with community involvement. Time and time again the standing principles of the VFC varied with the government policies. When they re-organised the VFC in accordance with the government guidelines they realised that they had not been given any rights and concessions to utilise forests for themselves. For felling a tree on individually-owned land, a felling permit from the DFO was then required. Similar restrictions on the sale of certain MFP and royalty charges were other points for which the people were disheartened. For a tree trunk (above 5 ft from ground) measuring 35 inches in perimeter one was to pay a royalty of Rs.3.40, and for a quintal of *mahua* Rs.7.35.

However, people became encouraged with the new policy package of VSS offering more rights and concessions. The villagers decided to transfer the management of forests into the hands of women who perform regular activities in the forest. Initially, there were two different arguments favouring the transfer of management aspects to women. Most of the male villagers were convinced that women formed a group of regular visitors into the forest for the collection of fuelwood and MFP. Women also recognised how they utilise and exploit the forests in many ways. The second argument was that women, being regular visitors into the forests, could keep a watch over the forest and, once empowered, would apply their considerable knowledge for better protection, based upon their experiences. Some conservative men who were in the beginning opposed to the handling of VSS by women agreed ultimately with the others. In the next stage of action, the Mahila Samiti was given the VSS management and the VSS was constituted as per the guidelines on 30 May 1994.

Finally, the action plan at the village level was prepared by women. Night patrolling in the forest was agreed to be operated by two male persons posted at either ends of the forest which communicates to roads or cart-tracts. But the wood cutting and smuggling could not be stopped. On seeing the problem, the president of the VSS decided to employ a remunerated person to patrol in the forest regularly and instead of guarding one end of the forest, he would roam around the forest. It was further decided to pay remuneration in kind (6 quintal of paddy per year) with equal contributions from each household. After one year, the person's contract with the VSS was to be terminated and a new person would be appointed. The village youth were motivated to provide free service in the protection of the forest. All men in the village promised to help in preventing illegal wood cutting. The collection of dry leaves would be done by one person from each family daily to meet household requirements and no family would be allowed to send more than one person.

A fine of Rs.200 was to be charged to a villager for cutting a tree without permission of the VSS. For outsiders, the penalty could be much higher than Rs.200. The penalty and half of the amount from the auctioned seized wood went to the community fund. The other half of the auction money was allocated to the Forest Department.

The women who led the VSS wished to include some men in the committee who would decide disputes and supervise patrolling. All the general members of the village should also participate in meetings convened by the VSS. However, the management was confronted with problems that if the EC of VSS convened emergency meetings, women failed to attend the meeting, engaged elsewhere as wage earners or for household jobs. When the EC decided a visit to the forest to check the condition of the forest and prepare management strategies, not all members could join and some families sent young representatives who were not able to contribute. Sometimes differences in opinion appeared between the Executive Committee members and general members. However, these problems were not considered too grave and most of them were decided at the village level.

For a better management of the forests, the village people came up with some suggestions which, if they materialised, would strengthen the organisation.

- A trench line around the forest as a demarcation line might help to protect the forest.
- Afforestation on wastelands on the Forest Department's initiative and final benefit-sharing with the community; so that the plantation would provide income for the village people.
- Establishment of cottage industries where forest produce could be processed for ready marketing.
- Employment for local people should go along with plantation and forestry activities.
- Construction of a community hall.

For all these needs people expected substantial help from the Forest Department.

BUDHIKHAMARI FPC

Budhikhamari village near Betanati in Mayurbhanj District has drawn significant attention from the government, NGOs and the local public for its achievements in forest protection. During the 1970s the idea of forest protection was brought up by the people. Gorachand Mahant led the initiative for forest protection and is still the person behind the achievements that are to be seen today.

The forest that was brought under the management and protection by the villagers of Budhikhamari and adjoining villages is named the Manchabandha (loft-making) forest. The denomination indicates the past density of the forest. Roughly three decades ago, the forest was very dense, wild life was abundant and poachers were coming there and making lofts on large tree branches near water *ghats* for hunting.

Herds of elephants were coming from the Hatikut area to graze in the forest. During the rainy season, the elephant herds were remaining in the Manchabandha forest. Timber contractor's indiscriminate felling of trees then degraded the forests. The situation became so serious that the perennial water resources dried up, crops failed, the elephant herds left for other forests, and game animals disappeared. This resulted in a supply crisis of forest produce for domestic requirements. With this the socio-economic status of poor people who were depending exclusively on forest products for food and cash, greatly deteriorated. These situations provoked initiatives by Gorachand Mahant, who with the encouragement of K.C. Mishra (now Range Officer of Betanati), motivated the people to protect the forest. Their motivation was to relate the problem of deforestation to a socio-economic background, and the environmental utility of the forest to the sustained availability of forest resources. This motivation made people constitute a FPC in 1987. In the first motivation camp attended by The District Forest Officer (DFO) and Range Officer (RO), a verbal distribution of the forest among different villages for protection was done. The DFO gave proceeding registers to his FPC to keep regular records. On the other hand, however, the timber traders became more active and started smuggling timber whilst ignoring the FPC. In particular, they opposed G. Mahant.

One important incident experienced by G. Mahant deserves attention here. Once Mahant and his fellows seized a truck load of wood and informed the DFO of Baripada about it. The truck owner offered a bribe of Rs.1500 to Mr. Mahant but he didn't accept. The trucker then offered liquor to villagers of a nearby village who inhumanly battered G. Mahant and his 28 followers. The local police station was informed and the trucker was seized. Yet, when Mahant returned from the hospital, to his great surprise, a warrant of arrest was waiting for him citing that he took an offensive action against the trucker. Mr. Mahant was sent to jail immediately. After returning from jail, Mr. Mahant discussed with the DFO how to constitute a mobile patrolling team with 5 youth from each village who would be under the FPC of Budhikhamari. The DFO ceremonially inaugurated the mobile squad operation. This generated interest among 50-60 villages to become parts of Budhikhamari FPC. In 1988 the Budhikhamari VFC had 56 villages directly participating in the VFC, and 19 villages supporting the VFC. At present there are 90 villages belonging to Budhikhamari FPC.

In 1993 the Budhikhamari FPC was reorganised according to the official guidelines of the VSS. Forest protection with community involvement was done with the involvement of 90 villages and the total number of patrolling guards increased to 450. On a rotational basis 45 youth are engaged for patrolling everyday, and the Forest Department issues identity cards for each patrolling guard. With these identity cards the patrolling guards got a boost in their morale and felt themselves to be direct representatives of the Forest Department. Each patrolling guard was paid Rs.15 till 1996 and after that got Rs.25 for one day's patrolling, paid from the common funds of the respective villages. There is also a provision for a bonus to the guards when they detect smuggling or nab an illicit tree feller. Further, the Forest Department provided uniforms to 25 youth for which all the 450 guards attended interview. The uniformed boys and girls are considered the leaders of the patrolling teams. Now there are 12 lady members in the mobile squad and they accompany the boys and bear the responsibility to stop illegal cutting by women. All the patrolling guards are provided with a *lathi*, a cane stick to protect themselves. Utilising some funds from the social forestry project, some people have made plantations of Teak, Sissoo, Gambhari, Mango, Jackfruit, Black berry, and Eucalyptus on road sides. The following terms of protection and management have been agreed upon by the FPC members:

General procedures and regulations

- One headload of wood per family per week is permitted for domestic consumption.
- For additional requirements, a person must apply to the FPC president, who in turn sees if the person really requires the materials. The president, if convinced about the person's need, may allow 50 per cent of the requested material or may discard the application.
- In case of fairs, festivals and marriages, the FPC office-bearers supervise the collection of wood to avoid overexploitation. They also supervise silvicultural operations, and see that the beats are equally distributed among the participant families.
- Although the Forest Department has approved selective felling and rotation harvesting, the FPC is reluctant to fell a single tree. The Forest Department's position that without

rotation harvesting and selective felling the timber is not growing well receives no attention by the FPC and their members.

Operational procedures

- The executive committee of the FPC meets twice a year to discuss future plans. Emergency meetings may be convened any time.
- Every household must contribute Rs. 5 per month to the FPC fund.
- Penalty charges may be collected within the range of Rs.50 to Rs.200 for small-scale exploitation, and the collections can be seized. Big smuggling is reported to the Forest Department.
- Women who do illicit felling or exploitation are charged with a penalty between Rs.10 and Rs.50.
- A person may be evicted from the FPC general body if the responsibilities to the FPC are ignored or the norms of FPC are violated. He will also be denied the entitlement to utilise the forest for his family.

PURIKHUNTA FPC

Purikhunta is a village 7 km away from Betanati with 107 households and a population of approximately 700 people which are mostly Mahant families. The village forest comes under the Betanati range office of Baripada Forest Division. There is a women-led forest protection committee engaged in protection and management of the forests around the village. The villagers are well-known in the areas as traditional conservationists of the local forests and exhibit an inherent tradition of forest protection. The people hold that two decades ago there was dense forest around their village, but due to increasing commoditisation of forest produce, the neighbouring villages cleared the forests. They are happy to have saved some patches of forests with community involvement in the traditional way.

According to Parvati Mahant, the secretary of the women-led FPC, the FPC was formally established in 1993. Long before that, however, they had their own traditional village committee led by men who made efforts to protect the forests. Their neighbouring tribe, the Lodha, which is locally denominated as a so-called criminal tribe, caused many problems to the management of the forests cutting down old timber. In fact, they had been recruited by illegal timber traders for wages. The local people never opposed the Lodha for fear of being physically assaulted or murdered, as the Lodha only fear court trials and police interventions. Once it so happened that a woman complained to the local police for having been molested by a Lodha. The Lodha was arrested as a warning to other Lodhas not to violate women. This plead then made the villagers of Purikhunta hand over the management of the FPC to women (Mahila Samiti). Lodhas usually cut trees in the day time when women work in the forests, so this was seen as the best arrangement to safeguard valuable trees.

On the 7th June 1993 the FPC was constituted named *Indira Jungle Suraksha Mahila Sangha* in which a total number of 125 female members were mobilised. The villagers of Purikhunta, Ramachandrapur and Purnachandrapur planned to protect about 100 acres of Undemarcated Protected Forest under Purikhunta FPC. The organisation fixed the following principles for the effective operation of the FPC:

- The executive body and the general body must meet once a month.
- Women and men in groups should conduct patrolling during the day and night time respectively.
- Person(s) found cutting wood inside the forest without permission of the EC will be penalised with Rs.51.
- Members failing to attend monthly meetings must pay Rs.10 as a penalty. This is also applicable for a person neglecting patrolling.
- Disputes are to be solved at the village level first and then, if required, to be referred onwards.
- There will be no restriction on the collection of dry leaves and dry branches. By order of the EC of FPC the forest floor must always be clean. Otherwise the Lodha may set fire to dry leaves and eventually cause a big forest fire.

- Cutting of trees and poles from the forest is strictly prohibited.
- There is no restriction on the collection of MFP, mushrooms, tubers, edible leaves, lac, resin, medicinal plants and seasonal fruits.
- For festivals and marriages fuelwood may be collected from Dahi (wasteland).
- Emergency meetings must be attended by all members.
- Collections from mid-term forest clearings are to be equally distributed among all members.

Problems and suggestions

FOREST AREA DEMARCATION

Political conflicts and exploitation by neighbouring villages often create tensions in the implementation processes of any FPC. An FPC exercises complete control over a forest patch and therefore needs to know its territory of operation. Due to a lack of demarcation of assigned forests, an FPC sometimes fails to check exploitation by intruders, particularly by neighbouring villagers who claim their demand on the same patch of forest. Each FPC and the community represented by it must know their area, not only by a pillar-marking of the Forest Department, but has also by a trench line demarcation. A fence of thorny plant species may be made instead of a trench line which would be less labour intensive.

PLANTATION ON WASTELANDS

To meet the fuelwood shortage of the local communities as well as for employment and creating income opportunities extensive plantation must be carried out. For the species selection the respective FPCs may be consulted. Economic plantations like Eucalyptus (to meet the demand of paper factories) and commercial plantations of teak wood (*Tectona grandis*) may be a joint effort by the Forest Department and local FPCs with entire or partial usufructuary rights for the communities involved. Cash crops like Sisal (*Agave* sp.) may be planted as well as measures taken against soil erosion wherever demanded.

COTTAGE INDUSTRIES

It is evident that there is a need to establish cottage industries to process forest resources. Vocational training centres should train the local communities to improve their household economies by shifting their activities from the sale of crude collections to the processing of commodities.

INVOLVEMENT OF FOREST OFFICIALS

Local FPCs understand that they are empowered by the Forest Department for forest protection and management. Forest Officials are involved in FPCs and must attend FPC meetings at regular intervals; intervene in intra-village conflicts regarding the distribution of forest produce; inform people about the outcome of smuggling cases; check procurement of nationalised forest produce by non-licensed parties; and, above all, help the people to keep records of their step-wise development.

DEMONSTRATION AND MOTIVATION

The Forest Department must take initiatives in organising demonstration camps and field visits to make local people understand what a proper forest management can be. During the demonstration and field visits examples of cultural factors and local institutions of various communities must be duly referred to, which helps in forest protection and management.

SECURITY

Sometimes people feel helpless to halt intruding woodcutters. The patrolling guards sometimes do not like to report them for legal offences. They need a legal security mechanism with the presence of an armed forest guard (as Purikhunta women pointed out), or by an authorisation from the Forest Department to employ uniformed patrolling guards (like in Budhikhamari) to be legally empowered to counteract offences.

WOMEN'S PARTICIPATION

While planning protection and management measures for forests with the involvement of the local community or forest dwellers, women should be involved in the programme who are the major collectors of MFP.

Concluding remarks

With the deforestation and scant consideration of alternative sources of livelihood, forest dwellers have failed to maintain their old 'preserve and use method'. Government control over forest resources has created a sense of 'being deprived' among the local communities which paved the way for grabbing of whatever was available in the forest. Forest dweller's economic, social and cultural practices are disrupted through increasing deforestation and denudation. Indeed, a sense of identity crisis is looming large before them. In the present situation and without active participatory involvement of local communities in forest protection and management, neither the remaining natural forests can be preserved nor a high-yielding man-made forest be created. The case studies indicate that the motivation of local communities could be enhanced with the assurance of getting a share of the forest produce for themselves until alternative occupations are created for them. In the absence of any other remunerative occupations, wood-picking has becoming a trade venture in the forest zones, cutting across religion and caste boundaries.

Legal policies alone cannot stop woodcutting unless the woodcutters become wood growers. If a local community is entrusted with growing more fuel wood and other trees to meet their own needs and those of surrounding villages, with the assurance of a share in the produce, it can become the manager and protector of the forest. With adequate employment opportunities from integrated forest activities such as forest-based cottage industries, a sense of awareness about the impact of first receding forest area has been generated. Such awareness is reflected through the efforts of community protection of forests on government land. The promises which the JFM policy has made to the local communities sound appropriate in the present circumstances. Amendments to the policy incorporating important points from various cases must be added at the local or regional level. At this point in time, JFM seems to be the only means to conserve forests better than the established modes of protection.

Appendix

(Brief particulars of JFM Policy (Orissa State)

(A) *Resolution No. 10F (Pron). 47/88/17240/FFAH of 1.8.1988*

1. Rural community should be fully and actively involved in

protecting forests.

2. Villages will be assigned a specific role in the protection of Reserve Forests adjoining their villages.
3. Villages will be granted concessions in bonafide requirement of firewood and small timber under section 24 of the Orissa Forest Act, 1972.
4. The concerned DFO shall assign the peripheral Reserve Forests adjoining village/villages according to the compartment line and if it doesn't exist, natural boundary like nala, bridges, etc., will form the demarcation point. However, Reserve Forest area under section 18(1) of the Wild Life (Protection) Act, 1972 is excluded from the purview of assignment.
5. DFO shall constitute a FPC for each of the assigned villages. The committee shall comprise of any three persons out of the concerned *sarpanch*, ward members belonging to the village, the local forester, Revenue Inspector and VLW, as may be nominated by the concerned Gram Panchayat.
6. Total number of members in the committee shall be at least 8 including persons belonging to SC, ST, women and landless category.
7. *Sarpanch* of the Gram Panchayat as chairman and local forester as the convenor of the comittee.
8. Assigned villages will furnish an undertaking to the concerned DFO for maintenance and bear responsibility for fire protection and prevention of illicit felling, thefts of forest produce and encroachment in the assigned Reserve Forest areas.
9. The villages shall be entitled to obtain their bonafide requirement of small timber and firewood for household consumption only and not for sale or barter. However, the committee shall work out the individual requirement and supply of wood.
10. One can remove small timber and firewood from the assigned Reserve Forests on obtaining a permit from two authorised members of the committee.

11. The FPC will come into force with effect from dated 2.10.1988.

(B) *Resolution of - 10F (pron)-47/88/23638/FFAH of 13.10.1988*

1. FPC comprising the *sarpanch* of the concerned Gram Panchayat, the Ward Members belonging to the village and at the most three other persons of the village as the non-official members, the local forester, Revenue Inspector and the VLW as official members. Non-official members of the village level FPC may be selected by convening a meeting of the concerned villagers.

(C) *Letter No.10F (pron)-47/88/27328/FFAH of 14.12.88 from joint Secretary to Government of Orissa to the Principal Chief Conservator of Forest (PCCF), Orissa*

1. Villagers assigned the functions of protection and maintenance of the peripheral Reserve Forests would be entitled to obtain bonafide requirement of small timber and firewood for household consumption free of cost. They need not pay royalty as prescribed in the 'Rules on Scheduled of Rates for Forest Produces in Orissa, 1977'.

The scope of the 1 August 1988 Resolution was enlarged in December 1990 to include the protected forests for a variety of reasons, however, this scheme has not had the desired effect of generating peoples' participation in protection of forests.

Notification No.16700-10F (pron)-20/93-F& E dated 3.7.1993. The notification was issued on the subject of protection of reserved forests and protected forest areas and the enjoyment of certain usufructs by the community. Forest management has to be reoriented to forge an effective partnership between the Government Department and the concerned village communities. Decisions made to implement JFM in the state in which the Forest Department and the villagers of the adjacent village(s) will be equal partners in the tasks of regeneration and restoration of degraded forests.

1. The scheme will operate in degraded forest land (Reserved and Protected Forest, but not in Sanctuaries, National Parks, Commercial or Industrial Plantation Sites) with potential for regeneration in the areas where villagers are willing or can be motivated to offer active co-operation in regeneration

and protection of the forest tracts. A single village committee may be allowed to develop up to a maximum forest area of 200 hectare.

2. After a general meeting of all adults living in the selected village, the forest officer/s may recommend constitution of Vana Samrakshana Samiti (VSS) of the village by the Gram Panchayat. There should be one Samiti for a group of hamlets or villages if these are all located adjacent to the forest area, to be regenerated.
3. VSS should include an adult man and an adult woman from every household living in the village/cluster of villages.
4. Concerned Forester — the ex-officio secretary of the Samiti will convene meetings of the Samiti once in three months and keep minutes of the meetings.
5. Each VSS will form its executive committee for a term of two years at a time. Treasurer will be selected by the executive committee.
6. The committee will comprise 10 to 15 members. Naib Sarpanch as Chairperson, Ward Member(s) representing the village(s), 6-8 representatives including at least three women elected by the Samiti, concerned forester, concerned forest guard and a DFO selected NGO nominee of the area as the members. The concerned Range Officer will act as returning officer.
7. The forester and the forest guard will not have any voting right.

References

Society for Promotion of Wastelands Development (ed.), 1992, 'Joint Forest Management: Concept and opportunities', Protocol of the National Workshop held at Surajkund, August 1992. Society for Promotion of Wastelands Development, New Delhi.

Poffenberger, M., Bhattacharya, P.A. Khare, A. Rai, S.B. Roy, N. Singh, K. Singh, 1996, *Grassroots Forest Protection: Eastern Indian Experiences*, Centre for South East Asia Studies, University of California.

Table 1: FPC Case Studies

Sl. No.	Name of FPC	Village reference	No. of house-holds	Caste composition	Divisional Forest office & district	Denomination of forests	Area demarcation
1.	Mahapada		22	20 Kshatriya 2 Brahmanas	Dhenkanal	Rupabalia, 'B' Class RF North	75 Acre
2.	Chhatia	Paika Sahi-14 HH Majhi Sahi-12 HH Upara Sahi-30 HH Tala Sahi - 29 HH	85	Brahmana Kshatriya Khaira (SC) Sabar (ST) Juang (ST)	Dhenkanal	Rupabalia, 'B' Class RF	2100 Acre RF. 100 Acre Khesra
3.	Katalposi		50	Mahanta Majhi (ST)	Kendujhar	Balibandha , 'B' Class RF	50 HA. (including 15 ha Khesra Forest)
4.	Jamunalia		152	Munda (25) Gond (100) Mahakuda (25) (Milkman) Blacksmith (2)	Kendujhar	Village Forests Budhakamana (10 ha) Chhanakamana (25 ha) Masanipada (3 ha)	38 HA
5.	Budhikhamari	Association of 90 villages	×	×	Baripada Mayurbhanj	Manchabandha Forest RF and adjoining areas	
6.	Purikhunta	Association of 7 villages	107 (Purikh-unta)	Mahanta and others	Baripada Mayurbhanj	Purikhunta village Forest	Not demarcated

10

People's Institutions and Joint Forest Management

Possibilities and Problems in the Context of Rajasthan (India)

S.N. Bhise
A.S. Mehta

Introduction

THE State of Rajasthan with a geographical area of *c*. 342 000 sq km presents a picture of vast contrast as far as its climate, vegetation and population is concerned. There are areas where the rainfall is as little as 150-200 mm in the north and north-west part of the State where it is a great sandy desert, whereas in the south and south-east the rainfall is as high as 1000 mm. The Aravalli mountain range, 688 km in length, extends from the south-west to the north-east and divides the state into two unequal physiographical regions — the north-west portion of the State is a vast expanse of arid and semi-arid tracts, covering an area of 20 million hectares. The south-east portion of the state is comparatively fertile, but subject to soil and water erosion. Eight million hectares of land area is occupied by the Aravalli and Vindhya hill systems which once had a rich vegetation cover, but is now degraded. The notified forest area is 31,150 sq km, which is 9.32 per cent of the total geographical area. However, 3.83 per cent of the area has forest cover above 10 per cent crown density per hectare, and only 1 per cent of the area has a good forest cover of crown density of 40 per cent per hectare and above. Per capita forest cover in the State is 0.03 hectare which is far below the national average of 0.08 hectare. The state also has the largest population of cattle, sheep and camels. Livestock population in Rajasthan makes up approximately 10 per cent of India's total livestock, and contributes to 15 per cent of the state revenue. Livestock rearing is an important occupation for farmers to augment farm income. There is excessive pressure on the limited forest resources of the State. The vegetation varies from scrubby and thorny forests in the arid tract, to dry deciduous forests in the Aravallis and Vindhya hills, and remnants of semi-evergreen forests in the Mount Abu hills. The Aravallis and Vindhya Ranges are the abode of tribals who are poor and have marginal and unproductive land, on which they practice subsistence agriculture. They have become deprived of the natural resources on which their livelihood systems were dependent for so long due to severe degradation of resources and faulty government policies.

Historical background of forest management

In 1947, the year of Independence, the State of Rajasthan was formed with the integration of the 22 Princely States of Rajputana. Forests in the erstwhile Princely States were preserved by their rulers as game parks and shooting reserves. These reserves were the best available forests, as they enjoyed immunity due to traditional authority, lack of communication, and remoteness of the tract from commercial markets. The rest of the forests were under the *jagir* land system. In a *jagir*, property rights are transferred either temporarily or permanently to an individual subject of the State ruler. This person, the *jagirdar,* had full control over the village, its inhabitants and natural resources. Under the *jagirdar* the forests were well-stocked, local people had free access to the forests, and there were few restrictions on its usage. Shifting cultivation was prevalent and people had their own system of protection and management of forests. They met their requirements for timber, fodder and fuel from the nearby forests either free or in lieu of certain royalty which was paid to the *jagirdar*. To sum up, one can say that the forests were abundant and the pressure of human and cattle populations was low in those days. In absence of regular working plans, the exploitation of forests was generally done on the basis of issuing permits and the collection of royalty resulting in selective and irregular felling. It was only towards the end of the nineteenth century that the need for forest conservation was felt when attempts were made by major states to regulate felling according to scientific guidelines. Tentative working schemes were prepared for some important *jagir* and government forests which were the basis of the exploitation of forests.

After Independence and with the constitution of the State of Rajasthan, a change in the ownership of land occurred. This era marked the beginning of government command over lands with the abolition of the Jagir system. With the promulgation of the Rajasthan Land Reforms and Resumption of Jagir (Amendment) Act in 1954, all uncultivated land under their control became vested in the State. The larger tracts were handed over to the Forest Department (FD) as Protected Forests, and the rest were kept under the overall supervision of the Revenue Department. During this period there was large-scale indiscriminate felling of forests, which resulted in the depletion of resources for its inhabitants.

Institutional arrangements for forest management in the Aravallis

Institutional arrangements in tribal communities were designed specifically for the purpose of social control and discipline, as regards the use, management and conflict-solving regarding use of common resources. In the initial years when the area was characterised by Princely State rule and *jagirdar*s had control over the resources, the Rajput rulers upheld the traditional institution of the village headman called *gameti* or *mukhia*. The position of headman is hereditary and is the highest and most powerful in the village. The authority of the *gameti* was the highest in village matters. In addition to his traditional role he had a few responsibilities which were assigned to him by the official administration. In cases of scattered villages with a number of hamlets, there were headmen at the hamlet level, and there used to be a Panchayat consisting of elders from each hamlet of the village. The institution of headman and Panchayat were responsible for maintaining norms of social control. There were also caste Panchayats pertaining to a particular caste, which were very strong and had great influence on their members. The verdict of these local institutions was final in many social matters.

The *mukhia* was responsible for maintaining and protecting the forest boundary of the village and the protection of the area from outsiders. In case of any imbalance in the event of breaking the law, the headman helped to recover the social order. In resolving conflicts, the provision of traditional or customary law was relied upon.

Provision of positive and negative sanctions traditionally defined for various categories of behaviour was resorted to regulate life in the group. Ex-communication, imposition of fines, compensation and feasting, were all made use of as instruments to enforce discipline in the village. When the State took over the control of forests, the village headmen were still the contact persons between the State and the local people, and headmen thus used to help the forest officials in the maintenance of forest boundaries, in the detection of cases of forest offence, and in extinguishing forest fires with the help of villagers. Except for these roles they were not involved in the management of forests. This institution of the headman of caste Panchayats is still prevalent in the villages. However, it is restricted to deal with cases of social or caste conflicts, rather than of forest management. The strength of the caste Panchayats and village headmen is still respected, even in times of Panchayati Raj and police institutions.

A great change in institutional arrangements came up in the management of the forests of Rajasthan for achieving the objectives of managing the forests for commercial exploitation. This change included a new legislation in the form of the Rajasthan Forest Act, 1953, and a new cadre of trained foresters were recruited to attain the objective of the scientific management of forests. This period marks the replacement of whatever was present in the name of local institutions and initiatives in the protection of forests before. Instead, the local people were increasingly involved in destroying the resources on which they were dependent for centuries. This was primarily because of the faulty practices and policies which alienated local people from their own resources. Even the introduction of the Panchyat Act in 1959 did not make any difference, and the Panchayat had no effective say in the forest activities which remained mainly in the State Forest Department.

Forest policies

1952 FOREST POLICY

The first forest policy after Independence was formulated in 1952 which stated that village communities should in no event be permitted to use forests at the cost of the 'national interest', which was defined as defence, communication and vital industries. The policy was in congruence with the high priority accorded to industrial development over environmental sustainability in the National Five Year Plans. The forests were demarcated and became State property. The process of alienation of people from the resource they had been using for centuries was initiated through the appropriation of control by the State. Extensive forest patches were cleared for agriculture, and forest land were perceived as a source of State revenue. Forestry meant the raising of commercially-valuable trees in order to get a sustained yield of timber in perpetuity. Exotic species were introduced to create man-made forests, and scientific forestry was equated with the raising of industrial tree plantations. The policy also favoured State control over other natural resources like pastures, rivers, tanks, and ponds. The management of these resources was fully transferred to the hands of government agencies and thus local people were gradually and systematically alienated from these resources. Thus, the common property resources (CPRs) which were a source of livelihood and sustenance to the rural poor progressively became the near exclusive property of the State. This is clearly evident in the changing terminology used by authorities in defining the people-forest relationship. The terminology changed from 'peoples rights' to 'concessions', and from 'concession' to 'privileges'. As a result of these developments the local people became antagonistic and subservient to the Forest Department officials. The encroachment of the state over forests and people's rights to use forest products was not unopposed, but due to its superior command and power over resources, the State was able to subordinate village people and their demands. Village people thus became accomplices in the destruction of forests in return for paltry wages that were paid to them by the forest contractors.

FOREST SETTLEMENT AND WORKING PLANS

From 1950 onwards, forests were put to scientific management through the process of forest settlement and the preparation of working plans. The process of forest settlement alienated the local people from the forest patches which they had been using traditionally. Under the settlement procedure, patches of forests were classified as one 'forest block', and in this block rights and concessions were granted to people on the basis of revenue village consideration. Thus, people of one hamlet who were using a particular forest patch could have a free access to another patch with which they had no association, and which was far from their habitation. This created a situation of conflict where the customary use boundary of one hamlet became the rightful property of another. This is one of the reasons why encroachments on forest lands have become pervasive and not resisted by the local communities. In absence of clear and meaningful boundaries authorised by the State, the ability of local institutions and traditional systems in the protection of forests has lost its legitimacy. Thus, forest management during this period witnessed a shift from local-level management for subsistence to scientific management for commercial interests which persisted until 1988.

NATIONAL COMMISSION ON AGRICULTURE, 1976, AND THE SOCIAL FORESTRY PHASE

As late as 1976, the National Commission on Agriculture (NCA) suggested a change from conservation-oriented forestry to production forestry. It advocated clear felling of valuable mixed forests and planting these areas with suitable fast-growing species yielding higher returns per unit area. The NCA also admitted in plan documents that no special programme were to be undertaken for the upliftment of the tribal economy, except that the forest development projects benefitted them indirectly by providing them with opportunities to earn wages. Under the NCA, the forests were to be managed exclusively for the purpose of timber production for industrial use. Thus the entire thrust of forestry during the first four decades after Independence was towards a high productivity forest system. Based on NCA recommendations, the government introduced the programme of Social Forestry on the supply side to fulfil the subsistence demands of forest-dependent communities. It had become obvious that if people's demands were not met, it would be impossible to protect forests. This was to be met by raising trees on the village common and on private land under the Social Forestry Programme. This was perceived as a programme that would release industrial forestry from social pressure. The NCA had concluded that free supply of forest produce to the rural population and their rights and privileges had brought destruction to the forests, and so it was thought necessary to reverse the process. It was recommended that people should produce what they consumed free of charge using village land for this purpose and thereby decreasing the pressure on forests.

The impact of past policies can neither be assessed in terms of achieving sustainability in increased production of wood, nor in checking the process of deforestation. Large tracts of mixed forests were harvested and converted into plantations. Most of these harvested areas did not regenerate, nor could the government accord protection to these areas from grazing and illicit cutting. This resulted in an absence of regeneration and degradation of the forest cover. The Social Forestry Programme did not take off well. In the initial years, it was well received only by the medium and large farmers, as they could invest capital in raising plantations for higher returns, deferred over a period of 5 to 10 years. Moreover, they had surplus land where they could take up the programme. On the other hand, the poor small and marginal farmers had no capital to invest in plantations, and they had practically no land where such an activity could be taken up. The poor farmers were also more vulnerable because of the long gestation period (high risk) of forest crops. Due to these factors, social forestry was not a rewarding proposition for small and marginal farmers. The premise of the policy-makers was that once the tree crop matured it would find its way to the village market. However, the big farmers, who were seeking higher cash incentives, delivered their products to urban markets and forest-based industries which were able to pay them a higher price. In the end, the Social Forestry Programme failed to realise its goals and ended up making rich villagers richer and poor ones poorer.

1988 FOREST POLICY — EFFORTS FOR A PARTICIPATORY APPROACH IN FORESTRY

The Forest Policy of 1988 is entirely different from the two previous policies of 1952 and the one enunciated by the NCA in 1976. The new policy marks a shift from totally State-controlled forest management to decentralised participatory and local-need-oriented co-management. The new policy stipulated that the requirements for fuelwood, fodder and small timber of the community residing in the vicinity was to be treated as the first claim on forest produce. According to this policy, the forests were not to be commercially exploited for meeting the demands of the industry, but were to be maintained for environmental reasons and for satisfying the minimum basic needs of the people, and strengthening tribal-forest linkages. It also stated that the primary task of all agencies responsible for forest management, including the Forest Development Corporation, should associate the tribal people closely in the protection, regeneration and development of forests, as well as to provide gainful employment to people living in and around forests. The policy also stressed the importance of NTFPs and stated that they should be protected, improved, and their production enhanced with due regards to the generation of employment and income. The policy also advocated the promotion of local institutions such as the tribal and labour co-operatives and government corporations to replace private contractors engaged in the task of forest regeneration.

To translate the new policy objectives into practice, the Ministry of Environment and Forests (MEOF), Government of India, issued a circular on 1 June 1990 to the Forest Secretaries of all States and Union Territories providing guidelines for the involvement of village communities and voluntary agencies in the regeneration of degraded forest lands through an arrangement commonly referred to as JFM. The key idea is that in return for protecting forest land, and helping in the regeneration and management of these land, people will be entitled to enjoy the usufruct rights over dead fuelwood, timber, NTFPs, as well as a significant share in the final harvest of forest produce. Under the JFM, Forest Department and local communities are expected to jointly manage forest land. In turn, both share the responsibilities and benefit. In this context, JFM was adopted as the institutional arrangement of the 1988 Forest Policy. With this came a shift in the forest management objectives at the State level, too. Forests were not to be commercially exploited, but were to be maintained to conserve soil and environment and to meet the subsistence requirements of the local people, who were to be involved in the protection and management of forests. Many States adopted these guidelines for a participatory approach to forest management. In pursuance of the central directives, the State of Rajasthan issued guidelines in March 1991 for the adoption of this participatory programme for the protection and development of degraded forests with the involvement of local communities. For the first time it specifies the rights of protecting communities over forest lands. It also speaks of the NGOs' contributions. Salient features of this participatory approach are:

(1) Constitution of village forest protection and management committees of villages who exercise rights over a particular patch of forest for meeting their demands for fuelwood,

fodder and timber. Membership is open to one adult member from each household. Access to forest land and usufructuary benefits will be given to those who become members of the FPC.

(2) These FPCs could be registered bodies under the Societies Act, or they could be simply registered with the local Deputy Conservator of Forests.

(3) An area of 50 hectares of degraded forest land can be allotted to each such FPC for development and protection. Natural Forests do not come under the purview of the JFM programme benefits.

(4) The beneficiaries are to be given usufructs like grasses and NTFPs free without having to share these products with the Forest Department. At the time of the final harvest, FPC members will get a 60 per cent share of the revenue earned from the harvest of planted trees and the government will get 40 per cent after deducting government expenditure for plantation. The FPC in turn will invest half of the share received in tree plantation.

(5) Rights and responsibilities are to be shared between both the people and Forest Department and an agreement defining the rights and responsibilities is to be made between the FPC and the FD.

(6) Financial arrangements for the development of forests can be made as (a) grant-in-aid from the NWDB, Government of India; (b) locally from financial institutions; (c) grants from industrial and social welfare institutions; (d) mobilisation of resources from the society; or (e) from the Forest Department's ongoing programmes.

(7) The role of NGOs has been envisaged as helping the village people in the formation of FPCs and acting as catalyst for communal action.

(8) No ownership or lease rights over the forest land will be given to the beneficiaries/voluntary agencies (NGOs).

(9) In cases of failure to protect the area from grazing, encroachment etc., the usufructuary benefits could be withdrawn.

The institutional arrangement for JFM involves the interaction of the State Forest Department with the community-level institutions, and the Forest Protection Committee (FPC). The third institution is the NGO which is to play a facilitative role. JFM is thus an instrument to empower forest-dwellers and increase their access to and control over the use and management of forest resources, which were once under the exclusive domain of the State. It also aims at improving their livelihood options by giving them employment opportunities within the village. More than six years have passed since the JFM approach in forest development and protection was introduced in Rajasthan. Major afforestation projects like the Aravalli Afforestation Project are operated according to the JFM principle.

After the initiation of the JFM programme in 1992, 1700 FPCs have been formed in Rajasthan so far which are protecting and managing about 1.50 lakh hectares of forest area. On the objective evaluation of their performance, the FD has found that 56 per cent of the committees are either very active or active. Most of the committees in the non-active category are those which have been recently formed (Kapoor *et al.*). This is a very encouraging figure as the evaluation is based not only on survival figures of the plantations, but also on the institutional capability of FPCs which is crucial to the success of this programme.

It is a fact that the FD has developed awareness about the benefits of participatory forestry in recent years, and its understanding and commitment to share control and responsibility with local people is increasing. Today, there is less doubt within the FD on the capacity of a community to undertake this vast work. This is reflected in the way the FD hands over charge to local people and institutions. However, little effort is made to ensure that the local community is defined by common interests and that the dominant elements in rural society are not cornering the benefits of participatory forestry. In other words, little investment is made in enhancing democratic functioning and equity within these institutions. Little is being done to cater to the needs of capacity enhancement of these FPCs, or to take up the emergent issues of forest protection such as the exclusion of non-members, the management of protected forests, conflict resolution, or the distribution of benefits and management of assets such as village funds.

Through its experience, Seva Mandir, a voluntary agency based in Udaipur, has embarked on implementing the programme in the field. Efforts have been made to analyse the outcome of the programme focused on community participation. As the JFM programme gains momentum, certain issues are emerging which need the attention of policy-makers so that necessary changes can be brought so that the JFM partners, the state departments and communities derive maximum benefits from this participatory approach in forest management.

Key issues in the JFM programme in Rajasthan

RESTRICTING JFM TO DEGRADED AND BARREN FORESTS

The resolution issued by the Rajasthan Government speaks of sharing and a participatory approach in degraded forest lands which are taken up for development and afforestation under JFM. This has become a stumbling block in the advancement of JFM policy. As natural forests have not been included in the purview of JFM, there is no provision for sharing the produce in cases where natural forests are protected by FPCs. This has limited the scope of JFM. The definition of degraded forests is interpreted differently by foresters. It depends greatly on their attitude and acceptance of the concept of participatory forestry itself. The preference according to which land is selected for development is —

(a) barren forest land,

(b) degraded forest land with a minimum of natural trees.

Even if such areas are allotted for development to FPCs, the foresters have developed an opinion that any natural trees standing in such areas will not be included in the sharing of benefits when the final harvest is made. The order also confirms this. The principle that FPCs can take the benefits of sharing forest produce only from those trees which have been planted is gaining ground in the minds of foresters. This makes a mockery of participatory forestry, when people who protect an area are denied benefits, but those who may degrade forests can lay claim to future benefits under the JFM scheme. Such interpretation of the JFM guidelines illustrate the ambivalence of forest officers towards participatory forestry and the deep identification with notions of possession and control on the part of the forest bureaucracy.

There are instances where people of a hamlet have been protecting small patches of forests in the vicinity of village, but when such a patch is taken up for protection and management under JFM, the right to do so is denied to them by foresters on the pretext of the abundant natural vegetation on the site. This works towards undermining the collective initiative of local institutions towards the protection of forests.

Case study: Mohandungri Village

Seva Mandir has been working for several years in the fields of water conservation and afforestation (private and communal) in this village. Enthusiastic through the policy statement on JFM, the NGO and village people submitted a proposal to the FD in 1993 for the development of degraded forest land in the vicinity of the village. A FPC was also formed. The proposal was turned down by the FD on the grounds that the entire revenue village comprising of 12 hamlets was not included in the FPC. The NGO staff started again — a new FPC with 500 members from 9 hamlets was organised and the fresh proposal was forwarded to the FD in 1995. The proposal was again rejected, this time giving any reasons. The NGO contacted the FD official through umpteen letters and requests. The hopes of the village people were again rejuvenated when, in 1996, higher officials of the FD visited the site proposed to be developed under JFM. The sanction to the proposal was now denied on the pretext of the site being 'inadequately degraded', the basis being a number of standing trees. A new 'appropriately degraded' site, adjacent to the previously proposed site, was proposed by the FD, and this degraded site was finally given formal recognition in 1997. Thus, a period of about three years elapsed simply in the negotiations to *start* JFM!

Meanwhile, the morale of the people and the NGO, as well as their faith in the FD, was crushed. The revenue village criterion for forming FPCs is, on the one hand, overprescriptive (it may not always be feasible to mobilise *all* the hamlets that make up a revenue village, and the relative importance of the patch of forest to be enclosed under JFM will vary for different hamlets); and, on the other hand, it has often been ignored by the FD itself. In this logic, rejecting a site as inadequately degraded would imply: (a) that village communities are not able to gauge when they need to protect a forest, or alternatively, that forest protection is not contingent on need or demand from the villagers; (b) that there is no value in conserving areas which are in somewhat good condition. Rather, forests should be brought to the point where degradation reaches desperate proportions (it may even be a signal to the people that they should go ahead and degrade forests in order to take up JFM!). The whole process seemed like a game between two institutions, where the FD perceived the NGO as its competitor, rather than as a facilitator/collaborator, and was reluctant to endanger its own stake by extending support to NGO-initiated FPCs.

PLANTING vs. REGENERATION

The study of the JFM order reveals that degraded forest patches are to be regenerated only by adopting plantations. Benefit-sharing at the time of the final harvest also speaks of planted trees. But the evidence so far is that degraded forest areas in many parts can be regenerated without plantation on the basis of protection and subsidiary silvicultural operations. Natural regeneration is a far more cost-effective model than the plantation model. The following example may illustrate this: Talai Forest Protection Committee proposed the development of a degraded forest area of nearly 90 hectares in two parts. Part A, comprised of 40 hectares which had a good root stock which could regenerate if protection was provided in the form of enclosing the area. No planting proposal was included. The estimated expenditure was to the tune of Rs. 2.00 lakh. In part B, 50 hectares of barren land with no root stock was proposed to be developed with afforestation activities under JFM. The first proposal has been rejected whereas the second proposal is under consideration for approval with an estimated expenditure is Rs. 5.60 lakh.

WEAK LEGAL FRAMEWORK OF JFM

As a concept, JFM has no legal backing. It is a mere order issued by the State Government, and can be withdrawn any day. This has implications for the future of FPCs formed and the benefits accruing from the thousands of hectares of forests they are protecting. JFM at present derives its legal legitimacy from the resolutions issued by each individual State Government. It does not have a statutory basis, which makes it easily reversible and creates an uncertainty in the tenurial rights of the protecting group. There is a case for strengthening the legal legitimacy of JFM orders.

(a) Confusion over boundaries: In Rajasthan, people of one or more revenue village/hamlets enjoy rights over the same patch of forest. All the inhabitants of these villages are provided rights of usage according to the settlement procedures, but on the ground, the traditional resource use boundaries of each village/hamlet are not well demarcated. As such, when a particular village wishes to develop forest land under JFM, the FD should recognise and demarcate its boundary in the forest, over which they have rights of protection and management, so that the village does not end up protecting the forest of some other village. This will also help to avoid disputes which arise due to an overlap of traditional and legal rights of usage.

(b) Absence of legal authority and legal recognition: The FPCs formed under JFM are supposed to take up the work of forest protection without any legal authority. In absence of any recognition by law the FPCs face certain hardships for protecting the forest area from illegal grazing, illicit felling and encroachments. There are instances of police action against FPC members, who sometimes pay heavy penalties as individuals rather than as a group. This underscores their faith in the institution.

Power has been given to FPCs to charge compensation for illegal grazing and felling. They are able to enforce these rules on their members, but not on outsiders or powerful offenders who break the norms. For the smallest forest offence, the FPC has to approach the respective officers for their settlement, which generally becomes a tedious process.

CHANGE NEEDED IN MARKETING AND ENHANCING INCOME

(a) Access to NTFPs: The JFM programme revolves round the fact that people would start getting the benefits from NTFPs and the free collection of grass. This has not impressed people to a great extent in Rajasthan. People were getting grass free of cost from forests prior to JFM. The only difference now is that they have a greater say in the distribution and management of the JFM area. In the case of NTFPs, people had access for free collection of most of NTFPs except those which are nationalised like *patta tendu* (leaves from which bidis, cigarettes are made). This has not changed even after the JFM resolution was issued. The other NTFPs (*mahua* flowers, herbs, etc.) collected are to be sold to LAMPS at the rates fixed by the Tribal Corporation and the FD, and not by market supply and demand. Thus, JFM has not brought about a great change in the scenario. The tribals are still exploited either by middlemen or in the form of low prices offered by government collection centres. It is essential that usufructuary rights should be extended to nationalised items like *patta tendu* and NTFP like bamboo to enhance the income of FPCs, and that the government should open markets for NTFP trading so that the collectors get fair and competitive prices. This would ultimately require a reassessment of the existing policies of nationalisation of certain NTFP.

(b) Developmental needs and JFM: The main objective of JFM is to improve the lot of rural people by involving them in the protection and management of forests and providing them with greater access to natural resources. Thus the programme cannot be seen in isolation from other development programmes, unless it is integrated with other rural development programmes.

In Seva Mandir's area of operation it became increasingly clear that protecting and developing degraded forest land is not people's first priority. One community displayed an interest in developing the overall natural resource base (land and water) of the village thereby securing their means of livelihood. Thus, we integrated the concept of JFM with combined works on watershed development, small lift irrigation schemes, pissiculture and so on, in and around Shyampura village. Most crucial is the fact that all these initiatives are carried forward by organising a village institution with a collective stake in the management of common property resources. There is a need to link JFM with combined development needs of the village which are generally non-forestry activities. This type of change in attitude will change the overall economic condition of the village people who will then have more interest in the protection and management of forests.

NEED FOR CHANGE OF SILVICULTURAL PRACTICES

The objectives of JFM are explicitly different from traditional forestry objectives. Experience has shown that wherever the community received immediate gains from JFM, FPCs are doing very well. People can be motivated to take up JFM in exchange for some immediate gains, whether in the form of grass, NTFP or something else.The agencies involved in the implementation of JFM will have to offer a technical package of afforestation which should include such trees, shrubs, grasses, which start giving benefits at the earliest possible stage. Other incentives, such as income-augmenting programmes for women (and most forest-dependent communities) would also enhance their stake in the programme.

MECHANISMS FOR CONFLICT RESOLUTION

The JFM guidelines have not focused on evolving mechanisms at the village level for resolving disputes which arise between communities during the process of implementation of JFM, or after. Conflict resolution is left to the implementing agency which in the majority of cases is the Forest Department. The dispute may have to do with membership rights in the FPC, forest boundary definitions, privileges and concessions on forest produce, traditional and legal rights over forest produce, inter-village conflict over the use of forests, and so on. Most of the disputes, if not tackled expeditiously and with the consensus of the stakeholders, result in protracted disputes and in the attrition of social capital. The role of stakeholders (in this case the village communities) in resolving the conflicts cannot be overemphasised, but there is also an important role for the implementing agency (whether FD or NGO) to mediate wherever necessary. The implementing agencies could help to establish a new constitutional order for forest protection and management based on the shared interests of multiple units of local associations. At present, it appears that the FD is knowingly disregarding these problems and using shortcuts to solve the disputes. Sometimes they simply do not address the problem and feel satisfied by not working in the village in the light of the conflict. This practice needs to change, and the FD has to take the initiative to work on conflict resolution together with the village people.

FOLLOW-UP OF THE SHORTCOMINGS OF THE RESOLUTION ISSUED BY THE STATE GOVERNMENT

The guidelines and resolutions issued by the State Department in connection with JFM are seen as legal documents and are taken in a very rigid framework. In a programme where people are involved and which is participatory in nature, certain flexibility would be desired in the approach of the executioners. During the very short period of working on participatory approaches, the State Forest Department, NGOs and people have been in a position to make suggestions that would further strengthen the programme. The experience of Seva Mandir and that of some studies conducted have shown that community control and management can work in villages which are small, homogeneous, and where people are dependent on resources from the forests. On the basis of this observation, it was suggested that the unit of formation of village institutions such as FPCs should be a hamlet instead of a revenue village, which is the present stipulation of the resolution. Similarly, there was another suggestion for including the income from bamboo in sharing the usufruct from JFM areas which is at present excluded. There is some evidence that government agencies want to take lessons from such experiences, and set-up an effective administrational body which can incorporate suggestions and make forestry programmes more participatory and effective.

INSTITUTIONAL CHANGE

(a) Institutional change in the FD

Since the adoption of the new forest policy, a number of initiatives have been taken to orient forest officials towards their new role in the participatory approach towards forest management. A lot still needs to be done in training them in extension and community development in order to turn them into effective community animators. The majority of Forest Officers are still averse to the idea of including NGOs as partners in the JFM programme. They might well have valid reasons for this, but then these reservations should be shared so as to make amendments to the guidelines. The JFM programme is still in the initial stage in the State. Getting the people's participation in a short period is not easy. There are so many things involved in the programme as regards protection, management, the sharing of produce, and inter-village conflicts, which will need attention. The FD cannot address these problems alone. Perhaps it is here that NGOs can share the responsibility in co-operating with the FD more on equal terms. The JFM guidelines should therefore mention with much more clarity the role which the NGOs are expected to play in JFM. In absence of a clear statement, the provision is interpreted in a loose manner, as a consequence of which only token involvement of NGOs is sought.

(b) People's institutions

There is a need for a long-term commitment to make village institutions (FPCs) truly representative, accountable and effective. Few efforts are being made for capability building of these FPCs. The FD should be aware of the fact that an association of 4-5 years with a FPC may not be sufficient to effectively raise its capacity to protect and manage a piece of forest land. If these institutions do not represent community interests, the JFM concept will fail like so many other schemes paying lip service to participation. The success of the JFM programme will depend upon the strength of FPCs, and researchers have shown that state support is important for providing a back up to the local institutions enforcing decisions.

The real challenge to the FPCs will come at the moment when the FD's financial support will cease to flow to the FPC, and it is expected to manage its affairs with its own funds. If the FPC is mature enough to take up the challenge, then the task of capacity-building of FPC can be taken as achieved. In addition, economic upliftment of the members needs to be done by undertaking activities in the village which augment their income (along with JFM) either individually or collectively. In this task, the lead will have to be taken by an agency involved in the actual formation of FPCs in the field — usually the State Forest Department or NGOs.

A village fund is an outcome of the voluntary contribution made by villagers for a common cause. Money collected in this fund may be derived from a variety of sources (e.g. from grass sale, contributions from individual members of the village). The JFM resolution does not take care of how the FPCs are going to protect the degraded areas under regeneration till maturity. This is especially relevant in Rajasthan where the gestation period for tree crops is quite long. The age-old traditional practice of protecting the forests by rotation (*Suiya* system) is very rarely practiced anymore. Most of the FPCs engage a watchman to protect the sites where the FD is carrying out plantation work. There is a provision by the FD in JFM projects to provide for a watchman for at least 5 years.

Under the watershed development programme executed under the Ministry of Rural Development, GOI, through the District Rural Development Agency (DRDA), the guidelines provide for compulsory contributions from the beneficiaries, which are a percentage of the total cost of the work being done. This amount is known as the village fund which is to be used by the village institution for the maintenance of assets. There is no such provision in the JFM resolution, but it has been adopted by the FD for some of the FPCs it has constituted. It is essential that such village funds are created at the FPC level so as to facilitate the maintenance of the asset in future. Our experience is that wherever we have been able to create such funds, the FPCs feel more autonomous and try to utilise these funds only in cases of necessity, and try to develop a system of replenishing the funds after use.

(c) The Institution of District and State-level working groups

The State of Rajasthan is yet to take action on the formation of District-level and State-level working groups in spite of the government orders to create hem. At these fora, problems relating to the programme can be discussed and responses can be evolved jointly by the FD, local communities and NGOs. There is no forum, however, where the people and organisations working in the field can come together to exchange their views of problems and experiences. It is high time that such groups are constituted to serve as effective consultative mechanisms by providing feedback on the people's perceptions of the programme to the implementors. This would also help in making the programme more participatory and effective in its execution.

(d) Networking of FPCs

Seva Mandir is recognised to have a tehsil level network of FPCs, and has encountered several issues which required institutional interaction at various levels. The presence of an NGO as an interface was felt essential. For a dialogue between the community and the FD. Thus Seva Mandir's efforts were directed at organising FPCs into a tehsil-level network. The role of networking is to evince institutional support from the FD through collective bargaining and negotiation. The network is being looked upon by member FPCs as a forum where members can voice their concerns and raise issues regarding JFM. The experience is rewarding. The networking has given momentum to this programme making the members more active and self-reliant. Ultimately Seva Mandir will have to think about an institution in the form of a federation of FPCs at the State level.

Impact and possibilities of JFM

The State of Rajasthan has been executing the JFM programme for the last 10 years only, but there has already been a great impact on the people as well as on the FD staff, even though many of the officials still have reservations. JFM in Rajasthan, as in the whole country, is at a crucial stage in its evolutionary phase which is likely to determine the future of forest management in the State and the rest of India. There are a lot of success stories where FPCs have protected and developed their resources very well and are getting benefits from the sharing of the produce. Eklingpura, Salukhera, Malpur, Gorela, Shyampura, are a few examples of village communities in southern Rajasthan who have built their village fund from the resources developed under JFM, and are utilising it for forest protection and management with the help of a strong village institution (FPC). Evidence from research conducted appears to indicate that where the JFM programme has been conceived with care and implemented with vigour, degraded forests have regenerated. This has provided many villagers with natural resources and economic benefits to meet their sustenance needs, and thus provided incentives to be involved in the protection of forest areas. FDs in West Bengal and elsewhere in India have shown that government agencies can develop resource management in agreement and through co-operation with communities where both parties benefit, and the results are a rapid increase in biomass, genetic diversity and forest productivity. JFM can be taken as one of several options of sustainable forest management, but should not be treated as a band-wagon. JFM needs judicious planning, repeated consultations with the concerned people, and institution building activities before the programme can be implemented. If the issues outlined here are addressed properly, there is hope that JFM will turn out to be an appropriate and effective management tool to regenerate and protect Rajasthan's degraded and natural forests.

References

Agarwal Chetan & Sushil Saigal, 1996, *Joint Forest Management in India — A brief review*. Delhi: Society for the Promotion of Wastelands Development.

Ghosh, Abhijit, 1996, *Sustainability of Joint Forest Management in India,* National Workshop on JFM Jaipur. (unpubl.)

Kapoor, R.P., Ghosh, Abhijit & Upadhyaya, A.K., 1996, *Joint Forest Management in Rajasthan*. (unpubl.)

Mehta, Ajay & Bhise, S.N., 1995, *Participatory Approach in Forest Management — Policy & Practise,* (Public hearing on environment and forests). (unpubl.)

Pandey, Deep N., 1996 *Beyond Vanishing Woods. Participatory Survival Options for Wildlife Forests and People*. Himanshu/CSD, New Delhi.

Sarin, Madhu, *Policy Goals & JFM Practise — An analysis of Institutional Arrangements and Outcome* (Second draft), (unpubl.)

Sahai, U.M., 1997, *Participatory Forest Management*. Voluntary Paper presented at the XIth World Forestry Congress, Antalya (Turkey).

Saxena, N.C., no date, *First draft report on Policy Analysis*. (unpubl.)

11

Agroforestry Development and Management through Human Resource Development at the Grassroot Level

Laj Pal Bharara

Introduction

FOR centuries, traditional forms of agroforestry have been practiced and evolved by farmers throughout the world. More recently, agroforestry research, development and training activities have been promoted by government agencies, universities, training institutions, NGOs and local organisations. Agroforestry systems, both traditional and introduced, are of great diversity and relevance in the densely populated areas of South Asia and Southeast Asia. In the arid and semi-arid regions, the most appropriate and simple definition of agroforestry is the use of trees in farming systems in which farmers ingeniously combine the production of crops, trees and livestock. Research, development, extension and training should play a vital role, with the active participation of farmers, in constantly examining and re-examining the hypothesis that agroforestry systems are ecologically and socio-economically sustainable.

Agroforestry systems have evolved with changing agro-ecological, socio-economic and demographic conditions, while some traditional practices have been weakened by the same dynamic factors. These factors have created problems of rural poverty, illiteracy, escalating human and livestock population, land-use conflicts, deforestation, soil and watershed degradation, in addition to various other localised conditions in the arid regions. To overcome or minimise these problems, agroforestry needs to be promoted for increased production, productivity, equity in benefit-sharing, and sustainable land management and conservation (Tejwani & Lai 1992).

Agroforestry development through human resource development at the grassroot level is still in its infancy in India. Above the grassroot level, various National and International Organisations/Agencies like UNDP, USAID, ICRAF, ICAR, agricultural universities, ICAR Institutes, etc., have promoted agroforestry projects for training, research and extension personnel in agroforestry. Recently, Krishi Vigyan Kendras under the ICAR, had started grassroot-level human resource development for agroforestry development. The status of agroforestry training and education is generally poor throughout the Asia and Pacific Region. The Asia-Pacific agroforestry profile reveals that training, research and extension is the strongest in the Philippines. Other countries like India and Thailand offer degrees and training courses in agroforestry. However, there is a general shortage of qualified researchers, development agents and trainers (Tejwani & Lai 1992).

Agroforestry land-use systems hold serious social and economic implications in the fragile ecosystem of arid areas. Ecological deterioration and accentuation of natural hazards affecting balanced land-use systems has been occurring in the region due to human and livestock pressure. Fallow periods are shortened and cultivation has been extended to areas unsuitable for cultivation. Communities with static farming technology face various problems of over-exploitation of natural resources in order to meet the increasing demands of food, fodder, fuel and water resources. This has resulted in low productivity, low income, poverty, deforestation and nomadic and semi-nomadic patterns of living associated with migration and pastoralism. This causes misuse of land-use systems involving the cultivation of sub-marginal lands, overgrazing of range lands, deforestation and the depletion of the wood cover.

Systematic attempts to identify appropriate agroforestry systems for different localities and improve them for higher productivity and better economic returns were seldom attempted until the early 1970s. Research on agroforestery and silvipastoral systems were initiated at the Central Arid Zone Research Institute (CAZRI) in Jodhpur, India, in the early seventies. Various other studies highlight agroforestry practices, but overlook vital social issues. (Malhotra 1981, 1984; Shankarnarayan 1984; Shankarnarayan *et al*. 1987; GOI 1987; Baumer 1991; Ballabh & Saxena 1991 & 1992; Bharara 1970, 1988, 1989, 1991, 1992; Jodha 1982, 1985, 1986; UNEP 1991 & 1993). Not much research has been done on the socio-economic aspects, issues and prospects of agroforestry land-use systems on arid lands. Empirical studies on farm trees, their significance, changes over time, relationship with the agrarian structure and farming systems are still lacking. Even well-known Indian village surveys done in the post-Independence period do not mention of village trees at all (Ballabh and Saxena 1992). The present paper discusses socio-economic aspects, issues and prospects of agroforestry land-use systems in dryland agriculture, with specific reference to the arid zone of Rajasthan.

Concept and importance

Agroforestry has a long tradition in India. The socio-religious fabric of people throughout the sub-continent is inter-woven to a great extent with raising, caring for, and respecting the trees which are integrated in the crop and livestock production systems under different agro-climatic conditions. Multi-tier crop combinations in different areas meet cash and household requirements. The *khejri* (*Prosopis cineraria*) tree and crop combinations in the hot arid regions meet needs for fodder, small timber and food. Arable farming has dominated dryland agriculture so far. It is only recently that alternative land-use systems have been receiving move attention in dryland agriculture. This is because these systems give stability to the resource-poor, small and marginal farmers. The cereal grains and pulse crops grown by dryland farmers are just enough to meet their family requirements. Fruit products, in contrast, invariably result in a marketable surplus and thus bring ready cash to dryland farmers.

Fodder is an important requirement of all dryland farmers. The pinch of fodder shortage is felt most during the drought years. Some fruit trees, apart from yielding fruits, also yield valuable leaves. These leaves constitute nutritious fodder for cattle, (e.g. *ber* leaves contain about 10 per cent crude protein and are sold at a high price, especially during drought years). Apart from leaves, the fruit trees also yield wood from the annual prunings which can be used as fuelwood.

Fruit trees also provide employment for the farmer and his family members during the off-season. The agroforestry system generates additional employment opportunities for rural people. Nursery raising, budding, pruning, harvesting and packing are a good source of employment for children, women and school drop-outs. It is estimated that in the next decade, at least 25 per cent of the rural population must be provided with alternative employment beyond traditional farming. This will be very important to check large-scale migration of rural people to urban areas in search of employment. Hence, agroforestry and the associated industries like processing, packing, canning and so on represent potential employment for rural people.

Social factors influencing agroforestry

Agroforestry is the product of not only the physical environment but also of a social and technological framework covering land ownership and tenancy; irrigation systems; farmer impliments; market; and the attitude of farmers. Hence, addressing technological problems alone will not be sufficient to ensure success for agroforestry development plans and package programmes. Simultaneous attention has to be given to social institutions which at the moment are seen to be an impediment to achieving sustainable agriculture.

The apparent increase of agroforestry also appears to be linked to the social changes that have occurred over the past thirty or forty years. Population increase, institutional changes affecting natural resource use, and the linkage to wider markets push for the intensification of production. However, the way in which agroforestry has come into own is a rather circuitous route. Initially, agroforestry appears to be a private response to the ecological degradation and/or disappearance of village common lands, the major traditional source of fuel and fodder. If sufficient quantities are no longer available from common lands then attention turns to tree resources on one's own land. Farmers also realise the direct benefits to crops from agroforestry and are encouraging *khejri* on that account. In effect, farmers are reinterpreting and integrating trees into their rain-fed farming to enhance their personal resource base and intensify the production of crops, fuel, fodder and timber. In the process, they are also decreasing the risk of degrading their own land.

Pastoralist groups influencing agroforestry

The role that the trees and shrubs of agroforestry systems can play in animal production in a given area is itself influenced by the role that animals play in society. There are various kinds of social groups concerned with animal production and these have implications for agroforestry. The groups indentified are:

THE PASTORAL GROUPS

These are groups where animal rearing is the main, or only, occupation. This group includes nomads, semi-nomads and transhumants. They are traditionally in close association with their animals, constantly searching for pasture. The pastoral people as a whole are considered particularly well-adapted to the use of their environment.

THE AGROPASTORAL GROUPS

These groups are dependent mainly on agriculture but some also practice animal husbandry. In periods of drought, they acquire livestock from nomadic pastoralists at cheaper rates than usual. The agropastoral groups have been increasing faster than the pastoral groups, and in many places, encourage desertification by continuously pushing cultivation towards the more arid areas that are less and less suitable for permanent agriculture. Many development projects are changing their economy to a mixed farming one.

THE MIXED FARMING GROUPS

These groups consist mainly of settled farmers who invest in more or less sophisticated forms of fodder production and the management of herds and flocks. There are several agroforestry technologies that could help to achieve relatively steady returns such as crops for fodder or small fodder banks.

THE CROP FARMER GROUPS

These groups are very similar to the previous ones, consisting of farmers who are deeply involved in crop production and have few animals. They get fodder from the agricultural by-products and it would be expected that agroforestry would concern them more in respect of the improvement of cropping practices than for animal production.

Associations and Interdependence

The basic outline of agroforestry systems are a number of associations and interdependencies between trees, field crops and animal husbandry, that are also linked to fuel and timber concerns. Trees, crops and animals are a tripod which farmers have adapted to meet the fluctuations in their environment, with trees serving as the stabilising element for both crops and livestock. The spread and development of this system also allows for the intensification of production on a long-term sustainable basis. The most important and widespread tree, perhaps as a result of human selection, is *Prosopis cineraria* locally called *khejri*.

The *khejri* is managed and harvested like a crop, the activities and sequencing of which are neatly integrated with farm operations, labour availability, animal and human needs, and natural cycles. First, the random spacing of trees presents no difficulty for animal-powered operations. Draught animals, usually bullocks and camels, are yoked one at a time and can plough straight tight furrows through a thick stand with ease. Secondly, bean-like pods are produced during the hot pre-monsoon season when vegetables are scarce. The young tender beans are a supplementary and free vegetable, particularly for the poor. Third, the harvest occurs after *kharif* is over. This creates an extra month to six weeks of employment and wage labour during the slack season, particularly where a second crop is not obtained. Fourth, the *khejri* is harvested by lopping off its leaf-bearing branches. These are dried and the leaves are beaten off. The leaves are harvested during November and December. Fifth, the branches are used as a domestic fuel and for other purposes such as firing bricks. This frees animal dung from being used as a fuel, allowing most to fertilise the soil. Sixth, management practices dovetail with agro-metereological cycles for the two cropping seasons. This is significant if a winter *rabi* crop is grown. The removal of the canopy opens the area to sun exposure during the cool winter season.

Most farmers leave a few branches to allow trees to spread. In fact, some farmers carefully shape their trees to make as full a canopy as possible, although this is highly variable. In any event an assured source of seeds is associated with religious practices. Shrines dedicated to field Gods are found on most farmer's fields and are always shaded by a *khejri* which is never lopped or cut. The associations and inter-dependencies found in this system have been apparently long since recognised by farmers. However, as a dominant practice it appears to be a relatively recent phenomenon, gaining greater recognition over the past thirty years.

A mature tree can bring several hundred rupees as firewood when there is a need for cash. Farmers practicing rain-fed agriculture report, however, that they now have more trees than before, and with to have more. They also report that they actively encourage and protect *khejri* seedlings that appear in their fields whereas they used to ignore or uproot them during field operations. The practice also appears more prevalent with those farmers who, for more intensive land-use and to shorten fallows, erect boundary walls to protect their crops from grazing animals. Tree seedlings are also protected from grazing that destroys or retards the reproduction of trees.

Socio-Economic Advantages

The advantages of agroforestry in dryland agriculture are as follows:

(i) .It yields an extended range of products, (viz. food, fodder, fuel, fibre, fruits, fertiliser, etc).

(ii) It helps in the utilisation of off-season precipitation. Most of the drylands are cropped for a single season, but if a perennial tree component is present it makes use of rain throughout the year.

(iii) Agroforestry in dryland agriculture helps soil and water conservation. The perennial component of the agroforestry system either on contours, or in some other shape, helps in reducing soil and water loss.

(iv) It imparts stability and results in risk reduction of the two or more components of the system, so that even if one fails, the other should give a harvest.

(v) The presence of trees in the agroforestry system gives extended management options for the farmers (i.e. the trees can be harvested either for food, fuel, poles or small timber as per farmers' requirements and market demands).

(vi) Agroforestry systems stabilise the soil temperature especially during summer months. Thus, it protects the soil micro-organisms. Trees also serve as shelter for cattle.

(vii) It helps in recycling nutrients from the lower soil horizon. The deep-rooted nature of most trees results in tapping nutrients from deeper soil layers and returning them to the plough layer through leaf litter.

(viii) During the off season time can be effectively used for harvesting, pruning, and marketing of the tree components.

(ix) Tree components in agroforestry systems results in improvement in the micro-climate of the whole system.

(x) Agroforestry results in improving and stabilising the income of the farmers.

PERCEPTION OF AGROFORESTRY SYSTEMS

Of the various types of rural production, agriculture, pastoralism and agro-pastoralism are the most predominant. Agriculture is followed by more than 60 per cent of the households of the Rajput, Peetal, Bishnoi and Rajpurohit communities, on the other hand. Pastoralism in Rajasthan is the main earning source of barely 5 per cent of the total households belonging to Raika community. Agro-pastoralism is most commonly practised by more than two-thirds of the households of these communities having a combination of agriculture and livestock rearing as the major sources of livelihood. Agriculture and pastoralism are more risk-prone, and need to be supplemented with other alternative farming types. The general preference is to have a multi-component system of farming types, (viz. agro-pastoral, agro-silvi-pastoral or agro-silvi-horticulture-cum-pastoral (Table 1)).

Table 1: People's perception towards the proportion of income and risks involved (percentages) in various agroforestry systems followed by the desert dwellers

Predominant agroforestry system	*Percentage Income from different components ofagroforestry system and risks uncovered during an average year*				
	Agriculture	*Pastoralism*	*Silviculture*	*Horticulture*	*Risk Uncovered*
Agriculture	65	-	-	-	35
Pastoralism	-	60	-	-	40
Agro-pastoralism	50	30	-	-	20

Pastoralism-cum-Agriculture	35	45	-	-	20
Agro-Silviculture	60	-	15	-	25
Agro-Silvi-pastoral	45	30	10	-	35
Agro-Horticulture	55	-	-	15	30
Silvi-Pastoral	-	45	25	-	20
Agro-Pastoral-cum-Silvi-Horticulture	40	35	5	10	10

Indigenous agroforestry and its revival

A BIOMASS PRODUCTION-ORIENTED STRATEGY

The number of trees, especially *khejri* and *ber* bush colonies (called *malta*) within the plots, and *matt* (shelterbelt) on the field borders, are components of indigenous agroforestry systems in the villages of the arid zone. They play an important role in providing fodder, fuel and other materials besides keeping part of the crop land under natural vegetation. *Khejri* trees are lopped for fodder and fuel every year after the end of the *khejri* crop season. *Ber* bushes with deep roots are cut (harvested) before planting the crops. They sprout again and grow with the crops without competition. They are again cut after the harvest of field crops. The fodder from *khejri* called *loong*, and from *ber* bush called *pala*, are available every year. They are less affected by rain conditions compared to crop-based, biomass. Hence they are also called desert farmer's 'natural insurance'. During the drought years *loong* and *pala* fetch high price as foodgrains. During good rain years, the crops yield both grain and fodder around *khejri* trees, and *ber* bush colonies are often higher than in other parts of the same field.

Trees and bushes have an impact on the micro-climate and they help in the accumulation of humus. Unlike trees and *ber* bush colonies within the field, the trees, bushes and shrubs constituting *matt* (shelterbelt) are not harvested, except for collecting small materials for fuel and thatching. Thus, agroforestry components perform several functions besides acting as renewable resources for fuel and fodder.

Complementary usage of perennials and annuals

Indigenous agroforestry clearly reflects the farmer's strategy for a higher and more stable production of biomass. Perennials like *khejri* trees and *ber* bushes, alongside facilitating better performance of crop, act as a stable source of biomass for the farmer. Table 2 presents biomass production figures from perennials and annuals including planted crops.

During poor rainfall years high quality feeds (*loong* and *pala*) play an important compensatory role in overall supplies of biomass (i.e. fodder). This is partly due to the lower sensitivity of these sources to fluctuations in rainfall, and partly due to greater efforts by the farmers to harness them during the poor rainfall years. During good rainfall years, by-products of crops contribute the largest share to the total biomass production. But due to labour shortages and a sense of complacency, biomass from agroforestry components is not fully harnessed. Not only *pala* and *loong* remain unharvested, but in many cases *bajra* (millet) stalks are left uncollected. There is greater emphasis on collecting fuel, fencing and thatching materials (including those from common property resources).

Table 2: Biomass yield during different rainfall years in an arid village of Jodhpur district

Biomass yield	*Drought year (1963-64)*	*Average year (1965-66)*	*Good year (1973-74)*
Rainfall (mm)	159	259	473
Fodder yield (kg/ha^{-1})			
Top feeds	377	318	342
(Loong + Pala)	(78 per cent pala)	(40 per cent pala)	(53 per cent pala)
Grass	85	250	1850
Crop by product	62	435	2160
Foodgrain (kg/ha^{-1})	7	108	615

Fuel/fencing			
Thatching materials:			
Crop by-product (cart load)	0	2	4
Others (Cart load)	2	2	2

Incentives

The increase in the number of *khejri* trees in crop fields is quite encouraging. For other components, such as *ber* bush colonies (*malta*) and shelterbelts (*matt*) the situation is discouraging, as they are rapidly disappearing (Jodha 1985). The current initiative of the farmers to promote trees in the fields is a response to a felt need for drought insurance, especially after *ber* bushes in the fields declined following tractorisation. The decline of common property resources (CPRs), in area and productivity, also encouraged the protection and growth of trees as a part of agroforestry projects and as an alternative source of biomass. This is partly due to the privatisation of products of seasonal CPRs. Hence, as CPRs decline, agroforestry may emerge as a possible option to partly compensate for the loss. Tractor cultivation discourages the provision of trees or bushes in crop fields. Mechanisation is of little direct help to the rural poor, and modern scientific agroforestry is still unknown to the farmer. There is little scope for effective public intervention, legal or otherwise, but some incentives to farmers might help.

AGROFORESTRY POLICY FORMULATION

Agroforestry policies are generally formulated to attain a specified goal, such as productivity per unit area over time; a reduction in the instability of production; or the socio-economic security of a family and rural community within a designated span of time. The major criteria of agroforestry policy formulation should always be in terms of people's particular needs, goals and resources. Such a policy should be politically acceptable, administratively feasible, economically sound, socially trusted and capable of modifications as conditions change. Individual trees in cultivated and uncultivated land should be marked on a map so that indiscriminate felling can be stopped. If possible, the age of the trees or shrubs could also be entered in a pass book for every *khatedar* in rural areas.

The consolidation of holdings, the Land Ceiling Act, and land tenurial ownership may help extend agroforestry. Agroforestry can be developed by providing some incentive to the new allottees for raising trees along with crop cultivation. Agroforestry would help in income stabilisation and provide additional returns per unit area over space and time.

One of the most important policy implications relates to the perceptions of various groups of people involved in agroforestry. On the one hand, the farmers themselves have clearly perceived the decline in supply of needed forest products in the common forest land. They have started responding to the situation by increasing the quantity of trees on their private farm land. However, many government officials and policy planners presume that the farmers lack awareness and expertise.

Traditional agroforestry systems suggest that people take an active interest in tree cultivation and protection, provided it strengthens their production systems. The main policy implications for arid areas is, therefore, to develop trust between the intervening agency and the people. How is this to be done? One solution to this question might be that people will participate in agroforestry, provided the programme is economically beneficial to them. By corollary, they will not participate in agroforestry, simply for the sake of the environment and society as a whole. Agroforestry should be based on economic considerations for enlarging farmers' subsistence base. This is a necessary condition for success. Growing trees on private agricultural wastelands (as opposed to good quality land which can support more labour intensive annual crops) is both a socially and economically desirable activity.

The function of agroforestry is precisely to look into problems of land-use systems involving both agriculture and forestry. This will need a change in our research approach. To begin with, there is a need to appreciate the fundamental characteristics inherent in all such systems: that they provide multiple outputs from the same unit of land and that interactions occur between constituent plant species, one of which at least is a woody perennial (tree or shrub). We need to establish programmes of investigations that encompass more complex space and time dimensions than those normally encountered in agriculture and forestry. The interaction of trees, agricultural crops, grasses, and/or animals, will address some of the following lines of research:

(i) shade tolerance of various agricultural species,

(ii) forest species which protect the soil but do not reduce the energy level of the understorey,

(iii) breeding programmes designed to obtain in both trees and crops those attributes and characteristics necessary for successful and efficient intercropping,

(iv) the plant (topfeed)-animal interactions and their management.

A strategy for agroforestry and animal husbandry-based regeneration of arid areas should include:

(i) Strict implementation of the Rajasthan Soil and Water Conservation Act.

(ii) Land-use planning according to its capability and need.

(iii) Stopping the allotment of denuded forest land for arable cultivation.

(iv) Prohibition of the carriage of fuelwood and encouragement of community and farm forestry.

(v) Banning free and uncontrolled grazing in areas which are highly erosion-prone, and where erosion has already destroyed the fertile top soil.

(vi) Formulating a sound forest and vegetation policy.

(vii) Providing incentives for protection and plantation of trees in farm lands.

(viii) Enforcing village Panchayat social legislation for the protection and plantation of trees in farm lands.

(ix) Encouraging people's participation by offering vocational literacy programmes.

(x) Making use of traditional knowledge, and raising the motivation of the farmers influencing the adoption of new technology for increasing resource productivity.

(xi) Promoting a 'desert care' code-of-conduct to highlight

environmental conservation strategies to be followed by the pastoralists during migration, at camping sites and while returning from migration.

This includes:

(a) The establishment of a pastoral network, of pastoralists, agro-pastoralists and a forum on grazing lands development.

(b) The encouragement of manageable units of livestock.

(c) Sustainable land-use criteria for grazing of livestock have to be established.

(d) Judicious watering patterns have to be followed.

(e) Maximum production with limited and diversified resources is to be outlined.

Conclusion

The future course of action for alleviating the misery of dryland farmers, and boosting the production and productivity of drylands, should envisage afforestation through agroforestry on the level of the watershed. The improvement and stability of agroforestry systems in drylands will not occur unless the government creates a dependable infrastructure for meeting farmers' input, credit and marketing needs. Voluntary agencies functioning in these areas can join hands in this exciting venture.

References

Bharara, L.P., 1970, 'Social aspects of farm forestry in arid zone', *Ann. Arid Zone*, 9 : 36-44.

Bharara, L.P., 1988, 'Social policy for desertification in Rajasthan', In. *Desertification Control*, 7 (3) : 26-28.

Bharara, L.P., 1989, 'Socio-economic aspects of livestock migration and pastoral nomadism in drought-prone areas of Rajasthan', In. *Livestock Economy of India*, Oxford and I.B.H. Publishing Co., New Delhi, pp. 89-112.

Bharara, L.P., 1991, Socio-economic and legal aspects of grazing lands with emphasis on livestock migration and pastoralism in arid zone of Rajasthan, India. IV International Rangelands Congress. Montpellier, France, Serial 415.

Bharara, L.P., 1992, *Population pressure and its impact on Arid Zone of Rajasthan. India. World Congress on Natural Hazard Reduction*, vol. I. Institution of Engineers (India), New Delhi, pp. 13-26.

Ballabh, Vishwa and N.C. Saxena, 1991, *Socio-economic aspects of tree growing by farmers in south Asia. Institute of Rural Management*, Anand and Oxford Forestry Research Institute, Oxford.

Ballabh, Vishwa and N.C. Saxena, 1992, *Farm Forestry — A Review of issues and prospects*, IRMA, Anand, India.

Baumer, Michel, 1991, 'Animal production, Agroforestry and Similar Techniques', *Agroforestry Abstracts*, 4 (4) Nairobi, Kenya.

Government of India, 1987, *Description and classification of wastelands*, National Wastelands Development Board, Ministry of Environment & Forests, New Delhi.

Jodha, N.S., 1982, Decline of common property resources and its implication for livestock farming in arid region of western Rajasthan : Micro-level evidence. Paper Presented at the workshop on : Problems and Prospects of Livestock Development, Institute of Development Studies, Jaipur.

Jodha, N.S., 1985, 'Population growth and the decline of common property resources in Rajasthan', *India Population and Development Review*, 2 : 247-64.

Jodha, N.S., 1986, 'Common property resources and rural poor in dry regions of India', *Economic and Political Weekly*, 21 (27).

Malhotra, S.P., 1981, Traditional Agroforestry practices in arid zone of Rajasthan, Lecture delivered at the Summer Institute on Agroforestry in Arid and Semi-Arid-Zones, CAZRI, Jodhpur.

Malhotra, S.P., H.S. Trivedi and Y.N. Mathur, 1985, *Agro-forestry a Judicious use of desert Eco-system by Man*, Division of Human Factor Studies, CAZRI, Jodhpur.

Pandey, Viash N., and Pankaj S. Jain, 1991, 'Village woodlots in Gujarat : An experiment in Social Forestry', Working paper, IRMA, Anand.

Shankarnarayan, K.A. (ed.), 1984, *Agroforestry in Arid and Semi-Arid zones*, Central Arid Zone Research Institute, Jodhpur.

Shankarnarayan, K.A., Harsh, L.N. and Kathju, S., 1987, 'Agroforestry in the arid zones of India', *Agroforestry Systems* 5: 69-88, W. Jumk Publishers, Dordrecht, the Netherlands.

Tejwani, K.G. and Lai, Chun K., 1992, *Asia-Pacific Agroforestry Profiles*, APAN Field Document No. 1. Bogor, Indonesia.

UNEP, 1991, *Desertification Control Bulletin*, No. 19.

UNEP, 1993, *Desertification Control Bulletin*, No. 22.

12

Community Forestry in the Terai
The Churia Forest Development Project in Nepal

Jürgen Fichtenau

The project philosophy

THE Churia Forest Development Project (ChFDP) is based on the philosophy that, given a conducive policy and programme implementation framework, local people, if organised into self-help groups, can shoulder the prime responsibility for the management of forest resources. The external agency should only facilitate, either directly, or preferably through intermediary institutions, village-level organisations to become more aware of the potentials and deficiencies in their immediate environment. Based on their own assessment of the situation, external intermediaries should only facilitate a process enabling the villagers to find appropriate solutions to their respective problems. In order to empower the beneficiaries and to strengthen 'civil society', ChFDP works through intermediary partner institutions, and the main focus of GTZ support is therefore the empowerment of local development agencies to act as facilitators.

These local development agencies are namely:

- District Forest Offices
- District Soil and Watershed Conservation Offices
- NGOs
- Village Development Committees
- District Development Committees

- Banks

CORE STRATEGIES

To increase the production of forest products.

To decrease the consumption of forest products.

To alleviate poverty by creating alternative opportunities for (non-forest) income.

PROJECT GOAL

The ecological imbalance of the Churia Forests will be stabilised for the benefit of the people in Siraha, Saptari and Udayapur districts.

PROJECT PURPOSE

The target group successfully implements integrated and sustainable forest management practices.

Guiding principles of ChFDP's forest strategy

In the international context, the ChFDP approach should be seen as recognition of the Rio declaration and Agenda 21 where the cornerstones of the sustainable management of natural resources were laid down. These principles are: social acceptance, economic viability, ecological stability.

All activities concerning the sustainable management of natural resources in the ChFDP area therefore should not negatively effect future generations. In general, natural resource management (NRM) practices, and especially forest management practices, should not only take into consideration the present needs of society, but also consider the fulfilment of the needs of future generation. A sustainable system can only exist if the consumption of forest products does not exceed the forest reproduction capacity. In the context of Nepal, the guiding principle for ChFDP's approach is the Forestry Master Plan and the Forest Law and subsequent by-laws, which are based upon a community forestry approach. In the context of the ChFDP, the guiding principle is an integrated and participatory approach. The assumption is that addressing forest-related issues alone will not help to create a sustainable NRM system. Since acute poverty is the main reason for the overuse of the forest, an integrated approach includes efforts to facilitate access to additional income through various income-generation measures and loans. All these measures are geared to release the pressure on the forest resources.

The project area

BIOPHYSICAL FEATURES

The project area covers the 3 districts of Siraha, Saptari and Udayapur in the south-eastern part of Nepal (Eastern Development Zone). The area of the three districts totals 4.650 sq km. 32 per cent of the total area is covered by forests (with crown cover of more than 10 per cent/hectare) and totals 1.457sq km. From south to north the area covers 5 agro-ecological zones: the Terai, Bhabar, Churia, Dun (Inner Terai) and Mahabarat. The elevation ranges from 70 m above sea level in the Terai to more than 2.720 m above sea level in the Mahabarat mountain chain. The fertile plains of the Terai in the south, which are part of the Gangetic Plain, were covered with forests in the past, but have been extensively cleared for agricultural purposes since the early 1970s. Geologically, these almost-flat Terai lowlands consist of fine-grained recent alluvial deposits with low erosion risk.

The northern area, dominated by the mountain chains of the Churia and Mahabarat, tectonically active, geologically dynamic and ecologically very fragile. The materials of the steep slopes on the Churia Range vary from unconsolidated sands, claystone and sandstones, to pebbles, boulders and conglomerate. These areas are classified as 'high erosion risk areas'. At lower altitudes, the natural climax vegetation is a mix of tropical hardwood dominated by *sal* (*Shorea robusta*), and in higher elevations (higher than 1.500 m) a significant amount of Chir Pine (*Pinus roxburghii*) can also be found.

The climate is humid sub-tropical with mean annual temperatures between 20 and 24°C and an annual precipitation of about 1.650 mm (of which 80 per cent falls during the monsoon period between June and September). Four main river systems divide the whole project area into 25 smaller subwatersheds. The main river systems are namely: Kamala, Sun Koshi, Trijuga and the Southern Border River Group.

Demographic, socio-economic and livestock features

The population in the project area is about 1.2 million people, of whom the majority resides in the Terai and Bhabar zones of Siraha and Saptari Districts. The average population growth rate is estimated to be 2.3 per cent per year. The northern district of Udayapur is characterised by a strong in-migration, whereas Siraha and Saptari in the south show a higher out-migration. Altogether, 220.000 households are located in the project area, out of which 75 per cent have less income than required to meet their minimum basic needs. The average annual household income in 1995 varied between Nrs. 11.600 in Udayapur and about Nrs. 24.000 in Siraha and Nrs. 23.600 in Saptari. The whole area is poverty-stricken with the lowest levels of income in northern Udayapur. The population is ethnically and caste-wise very fragmented and characterised by a very low literacy rate of some 32 per cent, which is lower than the country's average rate of 36 per cent. For more than 90 per cent of the households, agriculture is the main source of livelihood. The average farm size of 1.1 ha in the Terai districts, and 0.9 ha in Udayapur, is far too small to fulfil the food needs of an average 6-person household. Moreover, one-third of all households are either landless or nearly landless with an average of less than 0.2 ha per farm. The average forest area per capita ranges from 0.76 ha in Udayapur to 0.02 ha in the southern parts of Siraha and Saptari.

For these reasons, only 20 per cent of all households have sufficient food for the whole year. Further, 80 per cent of the population food derived from subsistence agriculture and is only sufficient for 7 months in Udayapur, 5 months in Siraha, and 4 months in Saptari. A large part of the male population is therefore migrating to India for several months per year to look for employment. This puts additional pressure on women in the project area, who carry the main load of work in agriculture and forestry-related activities. Women throughout the project area are disadvantaged. This is reflected in higher illiteracy rates, reduced access to markets, limited chances to inherit land, all of which renders them virtually landless. A core problem in the project area is the amount of unproductive livestock. An estimated 400.000 cattle, 120.000 buffaloes and nearly 300.000 goats are kept. Low productivity, lack of fodder and weak support services for these animals are characteristic throughout the project area.

The forest situation

The findings described in this chapter are based on a forest inventory, which was carried out in the spring of 1995 as part of a pre-feasibility study. The results given in this chapter refer to the forest situation of the whole project area and derive from 186 randomly-distributed forest inventory sample plots (relascope plots), and an additional 70 plots in so-called 'woodlots' outside the natural forest area located in the Terai near the Indian border. As mentioned earlier, the forest coverage of the project area totals approximately 146.000 hectares (for forest areas with a crown cover of more than 10 per cent). About 100.000 hectares are located in the Churia Range (parts of northern Siraha and Saptari, southern Udayapur and a strip north of the Dun valley). The remaining 46.000 hectares are located on the Mahabarat Range in northern Udayapur. The dominant species in terms of volume is *Shorea robusta*, which accounts for 50 per cent of the local tree species. Other major species are *Terminalia tomentosa* (8 per cent), *Pinus roxburghii* (4 per cent), *Schima wallichi* (3 per cent) and *Lagerstroemia* (3 per cent). The rest are other tropical soft and hardwood species. The mean stem volume of the 3 districts is 99 cum/ha. The highest volume can be found in Udayapur with 126 cum/ha, while the growing stock in Saptari and Siraha is only 41 and 36 cum/ha respectively. The differences in forest coverage become even more clear when looking at the total growing stock in the 3 districts which amounts to 14.8 Mio cum in Udayapur, 0.5 Mio cum in Siraha, and 0.6 Mio cum in Saptari. With regard to forest density, 43 per cent are classified as 'open forest' (crown cover 10-39 per cent), 45 per cent as 'medium dense forest' (crown cover 40-69 per cent), and 12 per cent are considered to be 'dense forest' (crown cover above 70 per cent).

Due to extensive logging and firewood collection, the mean number of trees per hectares is only 252. The lower diameter classes are only represented in sub-average numbers, which indicates a predominant removal of young trees for firewood (easy felling, cutting and transport of low diameter classes). However, a mean number of 2.085 saplings and poles per hectares (with diameters between 1 and 10 cm) indicate a high regeneration potential if the forests are protected from human and livestock interference. The annual mean increment is estimated to be 4 cum/ha in Udayapur forests. Given the lower productivity of highly degraded natural forests in Siraha and Saptari, the annual growth is estimated to be only 2 cum/ha. Small forest patches and woodlots outside the natural forest area, located in Siraha and Saptari, have been measured separately. Here the dominant species are bamboo, Sissoo (*Dalbergia sissoo*) and

Mango (*Mangifera indica*). The mean stem volume of these woodlots is 91 cum/ha, for which bamboo accounts for 32 cum/ha and Sissoo 18 cum/ha respectively. The total growing stock of all these woodlots together totals 409.000 cum.

Looking at the total biomass availability per ha (divided into stem biomass, branch biomass and foliage biomass), the results show that there are on average 121 tons of biomass per hectare. However, looking at the biomass distribution in the 3 districts (Siraha 37 tons, Saptari 50 tons, Udayapur 155 tons) it becomes obvious how difficult the supply of forest products in the two Terai districts has been since 1995.

To show the importance of the results of the forest inventory and to demonstrate the frame conditions under which ChFDP is working, these figures have to be projected onto the fuelwood and timber supply and demand scenario for the 3 districts. Based on present trends concerning population growth and deforestation rates (without taking into consideration severe counter-measures implemented by the project like fuelwood plantations, and the introduction of fuelwood-saving devices), the fuelwood deficit in 1996 is 89.000 tons per year, and will grow to 155.000 tons in the year 2001 and 293.000 tons in the year 2011 (Table 1).

Large timber deficits in the two Terai districts could be, at least theoretically, compensated by the timber supply from Udayapur District. The overall deficit in timber will probably start in 2010. This should not underrate the fact that a big deficit in timber supply already exists in Siraha and Saptari. It is difficult, however, to estimate how much timber from Udayapur reaches Siraha and Saptari Districts as only little is known about the distribution of timber in the area.

Community forestry and private forestry

PROBLEM STATEMENT

The Churia belt is divided into two parts: the southern flanks of the Churia Range located in the northern parts of Siraha and Saptari, and the remaining Churia forests of Udayapur. Whereas the first part consists of degraded former *sal* forests, the second part consists of well-stocked and mainly mature *sal* forests. The reason for the degradation of the southern Churia flanks are a high population pressure and rising demands for fuelwood and timber from as far

Table 1: Fuelwood supply-demand scenario

Year	*Udayapur*		*Siraha*		*Saptari*		*Total*		
	Supply tons/a	*Demand tons/a*	*Supply tons/a*	*Demand tons/a*	*Supply tons/a*	*Demand tons/a*	*Supply tons/a*	*Demand tons/a*	*Deficit tons/a*
1991	226,000	205,000	21,700	31,000	23,000	60,200	272,000	296,000	**24,000**
1996	213,000	240,000	20,000	35,300	21,700	68,600	255,000	344,000	**89,000**
2001	200,000	276,000	18,300	39,700	20,400	77,300	238,000	393,000	**155,000**
2006	186,000	314,000	16,600	44,450	19,100	86,400	222,000	445,000	**223,000**
2011	173,000	352,000	15,900	49,400	17,800	96,000	205,000	498,000	**293,000**

south as Bihar in India. There is timber and fuelwood smuggling on a large scale. Another serious problem are free-grazing animals, which prevent natural regeneration of the forest and cause soil erosion on the steep and rugged slopes of the Churia range. The very fragile Churia range consisting of sandstone is further destabilised by domestic grazing animals. Yet, the Churia belt itself is thinly populated, especially in the Udayapur part. The lack of population pressure and the absence of roads in this area have prevented a more rapid degradation of the existing well-stocked forests. The main problems for the project, with regard to the project's overall goal to improve the ecological balance in the Churia forests, are:

- Acute poverty in the project area, which causes over-use of forest products, especially firewood, as a source of income.
- High population growth rates and migration tendencies.
- Illegal encroachment of forest land.
- Firewood and timber smuggling to India on a large scale.
- An uneven distribution of forest areas (depleted areas in the south, well-stocked forests in the north).
- An uneven distribution of forest land among the communities living in the proximity of forest areas. Former forest users from distant areas are cut off from the forest resources. This leads to an uneven distribution of chances to generate income.
- The size of forests handed over to communities are either too small (people are still encroaching neighbouring state forests) or too big for one single community to manage and protect.
- Huge numbers of free-grazing domestic animals, which prevent the natural regeneration of forests.
- Aggressive rivers with flashfloods.
- An insufficient fodder basis for domestic animals in the existing forest area.
- A negative balance of fuelwood supply from existing forests.
- No indigenous forest management system is practiced in the project area.

- Poor quality forest management plans (operational plans), which do not favour sustainable forest management.
- Insufficient land for fuelwood plantation (space is needed for agricultural production).
- Lack of motivation and technical know-how of the line agency staff (DFO).
- Lack of social skills when organising user-groups.
- Absence of District Forest Development Plans.
- Lack of fuelwood-saving technologies (wastage of huge amounts of fuelwood due to open fire places and brick factories).

To address all these problems by the ChFDP, a forest strategy requires clearly-set targets and goals, following a precise vision of how to improve the unfavourable conditions, along with a sufficient budget to fulfil these targets, and well-run co-operating project partners. The main focus areas to address the burning problems in this context are:

1. Increase of forest production and vegetation cover.
2. Decrease of forest product consumption.
3. Generation of income (from forest and non-forest activities).
4. Equal distribution of forest products in the project area.

Experiences with community forestry

In line with Nepal's forest policy, ChFDP has started to promote the community forestry (CF) concept as a forest strategy in the project area. The promising results with CF in the orientation phase have been implemented on a larger scale in the first implementation phase. As of May 1998, altogether 207 forest user groups (FUGs) have been formed, out of which 125 groups received more than 21.000 hectares of forest land as Community Forests. These 125 FUGs consist of more than 22.000 households (Table 2). These are impressive figures which exceed the targets set for the ChFDP by a long way. Even considering the fact that these figures are based on information given by the DFOs (with a poor reliability of data), this is a very remarkable trend.

Table 2: Forest distribution per district and amount of community forests

District	*Total forest area in ha*	*Handed over forests in ha*	*Percentage of total forest area*	*Benefitting households*
Siraha	13.500	3.098	22.9	5.534
Saptari	14.500	2.371	16.4	2.542
Udayapur	116.000	15.784	13.6	14.100
Total	**144.000**	**21.253**	**14.8**	**22.176**

Presently, 2.9 million seedlings are being produced in the project area, which come mainly from private nurseries. 392 hectares of CF plantations have been established by the Forest and Soil Conservation Office over the past three years. Surprisingly, according to available GIS information in the project area, nearly 13.000 hectares of potential afforestation sites and a total of nearly 50.000 hectares of potential enrichment plantation sites are to be found in the ChFDP area. In terms of quality control of implemented activities it is necessary however to take a closer look at CF under Terai conditions, and analyse its strengths and weaknesses. After six years of testing CF in the project area some conclusions can be drawn.

Looking at the positive effects of handing over national forests to local communities, it has become obvious that:

- The transfer of responsibilities to local communities to manage their forests have created strong feelings of ownership, and results in better protection of the forest resources.
- Daily needs for fodder, fuelwood and timber can be more easily fulfilled (because they are now legal).
- There is greater awareness and sensitivity among the population towards forest-related issues.

Community forestry, being a central part of ChFDP's forest strategy, is incomplete without looking at its problems. Main areas of concern with CF are:

- Forest user groups (FUG) are often formed without considering social problems (e.g. not involving the poor, landless or lower castes, ignoring political problems). The ensuing conflicts within the groups are often counter-productive to the planned forest management practices.

- The size of forest areas handed over ignores demand and supply assessments. This either leads to areas which are too small to fulfil the user's needs for forest products, or too large to be protected and/or managed by one small community. There is no assessment of the standing volume of timber and the Annual Allowable Cut with respect to the needs of local people.
- Existing Forest Management Plans cannot be considered to be sustainable, because in general they do not consider and reflect the potential of the forest, or the demands of local people.
- If forests are handed over, the community is generally only allowed to harvest 'dead wood'. Thinning operations of 'green trees', which are often necessary, have to be first sanctioned by the DFO staff. This leads to the common practice of debarking trees, thus letting them die slowly in order to be able to harvest 'dead wood'.
- Protection of CFs can lead to the encroachment of neighbouring state forests.
- Insufficient or no technical advice is given to FUGs by the DFO staff (due to time constraints, lack of manpower, lack of motivation and technical skills).
- The problem of equal distribution of forest resources between nearby and distant users. This problem will become a burning issue since most forests in the southern Churia hills in Siraha and Saptari will soon be handed over to communities living close to the forests. This will cut off the access to the forest resources for users living further away in the southern parts of the districts.
- Urgently-needed CF plantations in the southern parts of Siraha and Saptari are hampered by a shortage of suitable community land (land is needed for rice production) and there are low survival rates for tree plantations because of free-grazing animals.

This list shows that the CF model, which was originally developed for the hill districts of Nepal, cannot be easily transferred to Terai conditions in the plains. Adaptation and refinements are therefore necessary. The major differences to the hill situation are:

1. Terai forests consist of valuable *sal* forests (resulting in a reluctance by the DFOs to hand these forests over to communities as well as very restrictive harvesting rules).
2. The uneven distribution of the forest area (concentration of forests in the north).
3. Social problems with regard to landownership, untouchables and landlessness. Terai communities are part of a migrating society (lack of ownership feeling and responsibility for their village and the environment).

Experiences with private forestry

Besides CF, the Private Forestry Strategy is the second pillar of the ChFDP. On the background of uneven forest distribution in the project area, the southern part of the East-West-Highway has been nearly depleted of all its forests due to land requirements for rice production. Promotion of forestry on private land as agroforestry or farm-forestry has been one of the parallel core activities of ChFDP. Tree planting on private land promotes an increase in biomass supply from areas outside the community forests, especially with regard to fodder, fuelwood and small-size timber. It is encouraging that a number of private plantations already exist in the project area indicating the high potential of private forests. The land for private forestry or agroforestry development includes mainly privately-owned land under homesteads, farmland, bunds (small earthen dams surrounding the rice fields), fallow land, abandoned agricultural land, and is a viable option to meet the needs of forest products by the local population. The main aim of the private forestry strategy is to strengthen the self-help initiatives and experiences of local people in initiating forestry on their private land, and to mobilise the small and medium farmers with land-holdings up to 8 hectares per household to participate. Two NGOs working in this programme have mobilised approximately 2.000 farmers so far, organised into 70 private forestry groups. These groups have created 130 hectares of private plantations and 1.8 Million seedlings were produced in private nurseries in Siraha and Saptari Districts annually. Although these figures show a high potential regarding the Private Forestry Strategy in the project area, efforts in this sector have to be further intensified in order to compensate the immense pressure on the Churia forests.

The subwatershed management approach

A pre-feasibility study conducted in 1996 proposed to focus activities in one subwatershed per district due to the geographical dimensions of the project area and the limited budget of the ChFDP. An integrated and holistic approach in these subwatersheds was envisaged which could later be transferred to other areas and attract financial assistance from the German *Kreditanstalt für Wiederaufbau*.

An Integrated Subwatershed Management Plan (ISWMP) had to be prepared in order to document the strategies for replication in other areas. In order to evaluate the capacity of the natural resources, and the demographic and social-economic situation, in the respective subwatersheds it was decided:

- To take aerial photos of the total area of the 3 subwatersheds (approx. 28.500 ha).
- To conduct detailed Participatory Rural Appraisals.
- To make a detailed forest inventory of the forest resources in the subwatersheds.
- To hire a consultant with a back-stopping function regarding the interpretation of aerial photos, forest inventory and forest management planning.

Infrared aerial photographs were taken in November 1996 (scale 1:20.000), and Rapid Rural Appraisals were conducted in selected VDCs of all three subwatersheds to get a first overview on the social potential of the area. Activities which were started right from the beginning were large-scale dyke constructions to prevent floods, and in some places so-called 'catchment ponds' for water conservation were constructed. Although ChFDP is still active with the organisational and managerial set-up for the subwatershed management, the assessment of the forest resources and the technical recommendations regarding their management have been completed. The following paragraphs are an account of the findings, developments and recommendations with regard to forestry issues in the subwatersheds.

Aerial photograph interpretation and forest inventory design

As mentioned earlier, infrared aerial photographs were taken of the three selected subwatersheds. Infrared films were used to distinguish different tree species and different cropping patterns. The experience with infrared films did not fulfil expectations. As the aerial photographs were taken in October and November, owing to favourable weather in this season, *sal* which blooms around April cannot be distinguished from other hardwood species in a mixed forest, but species like *Dalbergia sissoo* are easy to detect. Agricultural fields have also been harvested at that time and are difficult to be identified from the air.

An interpretation key for the photo interpretation was made in May 1997. It was necessary to find reference areas to define forest types, forest degradation, forest density classes and agricultural features. At the same time, a detailed forest inventory had to be made. The three subwatersheds comprise of a total forest area of approximately 17.000 ha, including 'potential forest' area, which per definition of the interpretation key is not forest in the strict sense. It was decided to first conduct a detailed forest inventory in already existing community forests in the subwatersheds. A Geographical Positioning System (GPS) survey of these CFs was conducted by the Project Support Unit (PSU) and showed the following sizes:

- Kong CF in Udayapur: 1.023 ha
- Malati CF in Saptari: 80 ha
- Ghurmi CF in Siraha: 952 ha

Total: 2.055 ha

A regional forest inventory was made in 1995, with a second inventory in 1997 in the remaining forests of the 3 subwatersheds (approximately 15.000 ha with a lesser amount of sampling plots). With regard to standing volume, there are significant differences in the 3 districts. The average standing volumes of both inventories are compared in Table 3:

Table 3: Growing stock in forests 1995 and 1997

District	*Growing stock in cum/ha assessed during a Regional Forest Inventory in 1995*	*Growing stock in cum/ha assessed during a Forest Inventory in 1997*
Siraha	36	23
Saptari	41	38
Udayapur	126	110

It is far too early to comment on these figures, because the CF area in Saptari only comprises of 80 ha and is degraded, whilst the CF forest area in Siraha is also highly degraded. For conclusive statements, the results of another forest inventory after that of 1997 needs to be carried in the remaining forest areas of the three subwatersheds.

Collection of socio-economic data

Up to this point, only technical aspects of forestry have been discussed, but the social dynamics with regard to the actors involved (mainly the local villagers) for the natural resource management are of crucial importance. No sustainable forest management can be achieved without the approval of the involved Forest Offices and the members of the concerned forest user groups.

In the autumn of 1996 the first RRAs were conducted to learn about the social structures, expectations and potentials of the local people residing in the VDCs of the three subwatersheds. For the preparation of an improved Forest Management Plan, it was essential to collect data on households; numbers of livestock; the consumption of fuelwood; fodder and timber; forest management practices, and so on.

Socio-economic features of three CFUGs in three subwatersheds

The following is an overview of the most significant findings regarding the socio-economic features and requirements for forest products in the three CFs.

TIMBER DEMAND

According to the socio-economic survey, both the Ghurmidanda and the Malati CFUGs have no mature *sal* trees in their community forests. There are no other mature trees, or timber stands are inaccessible. CFUGs have to get construction wood from outside the community forests. Only Kong community forest still has good stands of *sal* from which the users can take timber.

FIREWOOD DEMAND

The average firewood consumption per household is given in the following Table 4:

Table 4: Annual fuelwood demand of 3 CFUGs

CFUG	*Number of households*	*Fuelwood consumption (average per household in kg/year)*
Ghurmi	474	2000
Malati	139	6400
Kong	418	7200

Source: New ERA 1997/Schnurr 1998.

LIVESTOCK

Grazing or the use of grass and leaves for livestock is a very important form of forest use in the Terai, and has great influence on the condition of the forests besides timber and firewood production. Natural regeneration of the forest is hampered by free-grazing animals and forest fires, which are laid by the herders to favour the growth of grass. There is a loss of fertile topsoil through run-off due to compression by cattle grazing in the steep parts of the forests.

There are several options for an ecologically more appropriate feeding of the livestock:

- using pastures (grazing land).
- using the remnants of agricultural crops.
- collection of leaves (from trees, bushes and bamboo).
- cultivation of fodder-grass (e.g. Napier).
- cutting of grass in the forest.
- taking the animals into the forest for free-grazing.
- procurement of fodder (e.g. straw, bran, oil cake husk, flowers).

The following Table 5 gives an overview on the livestock situation in the investigated CFUGs:

Table 5: Livestock situation in 3 CFUGs

CFUG	*Number of households*	*Total population*	*Number of livestock units*	*Agricultural land of user-group (ha)*
Ghurmidanda	474	2514	2341	355
Malati	139	796	1106	48
Kong	418	2507	2040	259

Source: New ERA 1997.

Based on investigations from comparable regions in India, the average fodder demand for one livestock unit is about 4.5 kg/day. The maximum fodder production from crop residues is estimated to be 1.28 metric tons per hectares of agricultural land, for highly productive areas, and 2.5 metric tons/hectares for pasture. The socio-economic study showed that in the Ghurmidanda and Kong CFUGs, besides stall-feeding with crop residues, free-forest-grazing is the normal practice. Grazing in the forest is allowed in both Operational Plans as there are no available pastures. According to the figures given above, the following demand/supply scenario was calculated for the two CFUGs:

Table 6: Fodder scenario in Ghurmidanda and Kong CFUGs

CFUG	*Fodder demand*	*Fodder supply from agricultural land*	*Uncovered demand satisfied mainly by free-forest-grazing*
Ghurmidanda	3845 tons/year	454 tons/year	3391 tons/year
Kong	3351 tons/year	332 tons/year	3019 tons/year

This example shows that enormous damage is caused to the forest by the grazing of livestock. However, the calculation does not take into account that the forest is also illegally used by cattle owners who are not members of these CFUGs.

A solution to the problem of grazing might be the system applied by the Malati CFUG, where livestock rearing is the main occupation of the households. As there is no grazing land in this CFUG, and grazing is strictly prohibited on community forest land, all domestic animals are stall-fed throughout the year. The users have cultivated

Napier grass and fast-growing fodder tree species in a forest plantation of 12.2 hectares. Additionally, the CF comprises an area of 77.4 hectares of degraded natural forest with extensive grass cover which is also used regularly for cutting. Based on the information supplied by the users, the following demand/supply ratio can be derived (Table 7):

Table 7: Fodder demand-supply scenario in Malati CF

CFUG	*Fodder demand*	*Fodder supply from agricultural land*	*Fodder procured*	*Fodder from cultivated grass and fodder trees*
Malati	1817 tons/year	61 tons/year	439 tons/year	1317 tons/year

In Malati, where improved breeds for milk production were introduced and stall-feeding is strictly applied, the available forest area of only 87.6 hectares can provide three quarters of the required fodder. At the same time, the soil degradation has been severely reduced by stabilising the soil through a permanent vegetation cover. A further improvement of the forest management could lead to a complete balance between fodder demand and supply in this CFUG.

Conclusions and recommendations

With regard to the management of forest resources in the project area the future strategy will have to address the geographical distribution pattern, forest policy, institutional set-up and technical forest aspects.

GEOGRAPHICAL DISTRIBUTION PATTERNS

Distant and proximate users

Traditional users of the once dense forest areas in the Terai have been cut off from their forest resources. People who migrated from the hill areas in the north and technically 'illegally' took possession of the forest areas in the Churia Range are now 'rewarded' by forest resources handed over to them as community forests. They claim that these forests belong to them, denying access rights to the more distant users, claiming that resources have become scarce. It is common practice, however, that users living in the area around the East-West-Highway (distant users class 1) either collect firewood and timber illegally in the nearest national forest, sometimes paying an 'entrance fee' to corrupt government forest guards; or use nearby community forests by paying a fee to the FUG.

A second group of distant users (class 2) lives near the Indian border, more than 20 km away from their forest resources. They mainly substitute firewood by using cow dung (which is lost as as fertiliser for their fields), or buy smaller amounts of fuelwood and timber in the East-West-Highway region. It is assumed that approximately 30 per cent of the forest products are smuggled or sold to India. A possible future scenario to solve these problems could be the integration of distant users (class 1) into existing FUGs by giving them access rights to the forest. Since they are not contributing directly to the protection or management of the forests, they shall have to pay higher prices for forest products. Furthermore, a strong support of distant users (class 2) would be helpful. Two main problems have to be tackled here, namely the lack of income to pay for fuelwood or timber, on the one hand; and the use of urgently needed fertilising cow dung as substitute for firewood, on the other. An additional negative side-effect of burning cow dung is that the toxic fumes cause lung diseases, especially for children. In these areas, therefore, access to loans and credits, as well as additional income generation measures have to be further promoted. An effective promotion of fuelwood-saving technologies such as biogas plants and improved stoves are also desirable.

The uneven distribution of forest resources in the project area

It has been deemed necessary and essential to create a regional institutional body whose task would be to take decisions concerning the appropriate management of forests in the three districts. This body would deal with the equal distribution of forest products in the 3 districts; benefit-sharing mechanisms within the project area; and to support and strengthen the development of local wood-based industries. Such a Forest Development Committee, or something similar, could be composed of representatives from all three districts, namely representatives from DDC, VDC, DFO, Department of Forests, FUGs, NGOs, wood-based interests and the ChFDP-PSU. One main task for such a committee could be to develop a forest product supply pattern in order to counterbalance the uneven distribution of forests and forest products in the project area. The richer forests in Udayapur could possibly provide excess forest products (from thinning and harvesting operations) to the poorer southern districts of Siraha and Saptari. A precondition, however, is the application of suitable forest management techniques. Furthermore, a Regional Forest Development Plan including the 3 districts has to be clearly developed stating the long-term vision for forest management practices, harvesting schemes, forest product distribution, and the development of suitable wood-based industries in the project region for the next 20 years.

This committee would also have to address the fact that forests in Udayapur district are ready to be harvested, whereas forests in Siraha and Saptari are mainly in the process of being improved. To allow them to grow to into similar dimensions would require enormous financial inputs. A sharing mechanism would also be desirable, which would enable certain funds accrued in the richer forests in Udayapur to be re-invested in the two southern districts. Otherwise it will be impossible to fulfil the rising demands for forest products (refer to Table 1) in this region, and further encroachment of the Udayapur forests, which already happens, will be inevitable in the future.

FOREST POLICY & INSTITUTIONAL SET-UP

The experiences of the project show that the CF model, originally designed for the hill regions of Nepal, cannot be easily transferred to the Terai without adaptations. Some of the main reasons are the unequal distribution of forest areas in many Terai areas, the abundance of valuable *sal* forests, and the reluctance of DFOs to prescribe harvesting operations in Operational Plans. Necessary changes have to take place at different levels. At the central, regional, district level and at the user-group level. Taking the project experiences into account, the following measures are proposed:

Central Level

According to the present rules and regulations, the total benefit from the use of community forests belongs solely to the concerned CFUG. This works well in the hills where forests are scattered more or less evenly across the country, but it is not suitable for Terai conditions, where the forest resources are very often (as in ChFDP's case) concentrated in certain areas. Community land for afforestations is scarce due to the fact that land is mainly privately owned by big landlords and is needed for rice production. It has to be realised at the forest policy level, however, that a sharing mechanism has to be developed, which both rewards the CFUG for managing their forests, but also creates a balance between poorer and richer areas. This does not only apply to forest product supply, but also to income equity. Furthermore, a tax on valuable forest products from the Terai should be introduced, from which Nepal's economy as a whole could benefit.

Another necessary change concerns the payment for services rendered for the preparation of forest management plans and technical assistance during their implementation. According to the existing rules and regulations, the DFO is obliged to provide these services free of charge. If this high quality work is provided by a DFO, then the CFUG should pay for the service. This would also be a good incentive for the concerned technical staff of a DFO (normally the ranger). However, the service should be seen as open for competition since the DFO's technical staff is quite often not in a good position to provide such services. In certain districts, especially where foreign donors are working, NGOs or other institutions are hired for the preparation of management plans. At present, however, it is illegal to pay for the services provided by the DFO. The preparation of technically sound forest management plans for the valuable forests of the Terai requires well-trained technical staff in the field of forest inventory and forest management. The Department of Forests should further promote education and training of its staff. The number of well-trained personnel in this regard is far too low, but could be raised by educating forest personnel in the Forest Campuses in Pokhara and Hetauda. The young generation of foresters generally has a sufficient theoretical background, but little practical experience, especially in technical issues such as forest inventory and management.

Regional level / District level

The decentralisation efforts of Nepal's government should be seriously taken into account when discussing the sound management of natural resources. This means that the DDCs should play a much stronger role in the future so that forestry-related matters are no larger looked after solely by the DFO. The above-mentioned Forest Development Committee and the development of a Regional Forest Development

Plan could be steps in this direction. Another crucial factor is the development of local wood-based industries in the districts. Once proper forest management is practiced, this will become a thriving industry. High quality forest management plans (Operational Plans) must include relevant data such as species composition, growing stock, and regeneration potential. Regional as well as detailed local forest inventories and socio-economic features have to be the basis for such management plans. The involvement of forest technicians other than the DFOs (either through NGOs or consultants) has to be discussed at the user group level, whereas regional inventories will remain the domain of the Department of Forest (i.e. the DFOs).

Another task at the district level would be a stronger co-operation with banks. In districts like Siraha and Saptari, where a considerable financial input is needed to create favourable conditions for the establishment of well-managed forests, long-term financial programmes will have to be developed together with banks. This should be the task of the DDCs, facilitated and supported by this project. With an increasing number of cases of conflict resolution in CFUGs it will become a more pressing issue than it already is. Political problems, social conflicts, non-functioning user group committees and the like, often hamper efficient natural resource management. DFO staff must be trained to assist communities to solve their problems in this respect. Finally, a strong CFUG network has to be established in order to share experiences and to learn from each other.

User group level

At the user group level more awareness has to be created with regard to the detrimental effects of open grazing and forest fires. The necessity for sound forest management, as well as the need for the integration of distant users into forest management, has to be thoroughly realised. Monitoring tools like a Project Book (a document stating all the planned activities, technical advice from the concerned LA office, accounts etc.) have been successfully introduced in user groups which implement soil and water conservation activities. The project book is a very useful tool to record the planned and ongoing activities. Similarly, public audits have proven to be of importance to keep track of financial expenditures, so that fraud can be much more easily avoided. Networking of FUGs should be promoted in order to create a forum where experiences and problems can be discussed. Here the FECOFUN (Federation of Community Forest User Groups in Nepal) can play an important role in co-operation with the DFOs.

FOREST TECHNICAL ASPECTS

In order to create a sustainable forest management system, clear management goals have to be set at the beginning of any management operation. Agreements have to be made at the start, stating how a certain forest area should be managed over the next 10 years or more. The forest management system to be applied depends on the requirements of the users and on the potential of the resource itself (degraded forest, mature forest, growing stock, and so on). To simplify the planning procedures with regard to different silvicultural treatments, required manpower and financial requirements, a forest area can be divided into management units. This can be either done by dividing the area into blocks of the same size, or into units which represent a homogenous type of forest (irrespective of size), making the application of a certain management technique easier. Workable compartments (containing one forest management type) should have a size of 20-50 ha with visible boundaries easily detectable, such as ridges, rivers, roads or paths. It is reasonable to differentiate between 7 different forest types (called Management Goal Types = MGT see Table 8), which represent different goals of forest management. The following MGT have been distinguished:

Table 8: Management goal types

Management goal type	*Main objective*
T 1	Timber Production (Plantation Forest)
T 2	Timber Production (Naturally grown Forest)
F 1	Firewood and Fodder Production (Plantation Forest)
F 2	Firewood utilisation (Naturally grown Forest)
P 1	Protection Forest — Complete Protection (majority of compartment consists of slopes steeper than 30 degrees)
P 2	Protection Forest — (50 per cent of the compartment consists of slopes steeper than 30 degrees)
RF	Reserved Forests — Timber production areas under the responsibility of the Forest Department (National Forests)

These different types of management goals are essential to meet demands for timber, fuelwood and fodder through the differentiation of plantations from naturally grown forests. In particular, different silvicultural treatments will have to be applied to take the steepness of the slope into consideration in order to prevent further erosion on the Churia Range. Finally, multiple management techniques will take into account the needs of national forests, which are located far away from villages, and are thus difficult to protect and manage.

References

AGEG-SIGNUM, *Prefeasibility Study — Churia Forest Development Project Nepal*, Main Report and Annex III, June 1996.

Chakraborty, R.N., Freier, Ines, Kegel, Friederike, Maescher, M., *Community Forestry in the Terai Region of Nepal*, 1997.

Fichtenau, J., Sherpa, S., Shrestha, M., *The Forestry Approach in Churia Forest Development Project*, 1998.

His Majesty's Government of Nepal, *Forest Act 1993*, Official English Translation, May 1995.

His Majesty's Government of Nepal, *Forest Regulation 1995*, Official English Translation, May 1995.

New ERA, *A Rapid Socio-Economic Appraisal of the Community Forestry User Groups in Ghurmidhar*, Mahuli and Kung-Hadiya-Siwai Subwatersheds of ChFDP, Nepal, 1997.

Poudyal, S.S., *Computer Printouts of Forest Inventory in ChFDP*, May 1997; Final Report on Forest Management Workshop with DFOs, Lahan, Sep. 1997.

Schnurr, Joachim, *Verbal Information and Draft Reports on Forest Inventory Design*, Interpretation Key for Aerial Photos and Community Forestry Model for the Terai, 1997 and 1998.

13

Participatory Management of Natural Resources with Forest User Groups

A Social-cultural Study in the Mid-hills of Nepal[1]

Christian Rosset
Klaus Seeland

Problems and background

FORESTS and the role of the people who depend directly and daily on their surrounding forest patches have changed considerably in Nepal over the last three decades. Scenarios developed during the 1970s once predicted the complete destruction[2] of the mid-hill forests in the kingdom of Nepal. The approach to the problem was focused mainly on ecological issues (the state of the forest) and forest users were mostly considered as the cause of the problem since the mid-hills were characterised by a growing population depending mainly on agriculture on terraces, with an increasing demand for firewood, farmland and grazing land.

Fortunately, these scenarios have not become reality. Considering many studies, Gilmour and Fisher (1991) even mention the likelihood that the forest area has not decreased for the last hundred years. However, the forest condition and, logically, the capacity of the forest

1. This paper is based on fieldwork conducted in Nepal by the Nepal-Swiss Community Forestry Project (NSCFP) in 1999.
2. A period of 15 years was claimed in a report of the World Bank published in 1978.

to satisfy the daily needs of villagers still remains a problem. The perception of this problem is, a decade later, radically different from that of the seventies. In particular, the significance of the social component of the forest has been increasingly acknowledged. Forest users are now widely claimed to be the appropriate actors to address these issues and their knowledge is taken into account. The Programme of Community Forestry (PCF) was started on these assumptions. The programme aims to transfer the right to manage national forests to the local communities who depend on them, namely the Forest Users' Groups (FUG). Hitherto *de facto* policemen, forest officials are increasingly adopting the role of an adviser.

In 1999, more than 7,800 FUGs had been recognised by the government. About 880,000 households were members of these FUGs (around 20 per cent of the households of Nepal) and 550,000 ha of forest were managed by them. The total forest cover of Nepal was estimated to be 5 million hectares. These figures reveal the significance PCF has gained since 1990 when the government seriously started to hand over national forests to FUGs. 1990 was a year of big changes in the Hindu kingdom as Nepal also became a democracy. Regarding the PCF, the changes were not only limited to the transfer of forest areas. A widespread democratisation process has also taken place. Before 1990, the representatives of the villages (called at that time 'panchayat village') were responsible for the management of local forests Post-1990, the executive committee of the FUG was elected by the members of the group, and the FUG no longer depends upon the Village Development Committee (VDC), the smallest administrative unit in Nepal.

In the overall context of democratisation and decentralisation, this paper aims to study the participation of members of FUGs in the Decision- Making Process (DMP) of the management of their Community Forest (CF) with particular respect to the participation of disadvantaged groups who have fewer opportunities to take advantage of the new situation. Two main research questions have guided this study. Which factors have an influence on the participation in the DMP within the FUG, with particular respect to disadvantaged people? What are villagers' and FUG members' main perceptions of the forests?

Objectives and framework

SPECIFIC RESEARCH QUESTIONS

The following research questions have been addressed over the course of this study:

- What are the formal and informal institutions of the FUG and their functions?
- What are the FUG members' perceptions of their CF?
- What interest do the members have to participate in the decision-making (DM) related to the management of natural resources, with particular regard to disadvantaged people?
- What values and expectations do the members have regarding the management of natural resources?
- How important is the CF for the members to satisfy their needs?
- How do the members participate in the DM related to the management of natural resource, and how do they express their interests?
- How do members influence, directly or indirectly, the decisions of the FUG? Why do they participate in that way?

FRAMEWORK AND DEFINITION OF THE KEY CONCEPTS

Fig. 1 presents the framework of the study. It introduces the elements that were determined as significant, considering the two main research questions. These elements will be explained in the next sections.

The policy arena represents the space where the participation

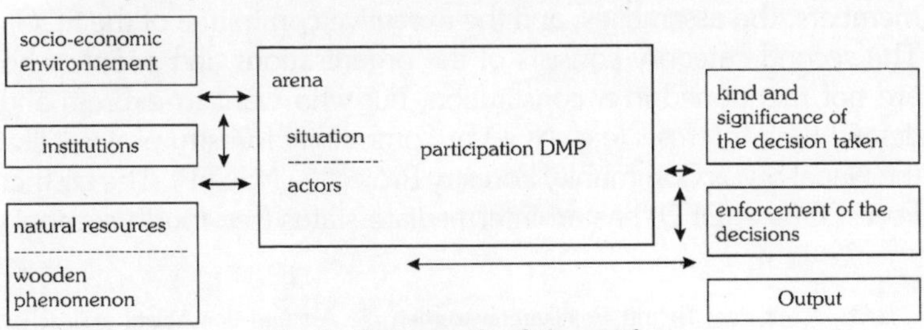

***Fig.* 1. Framework of the study.**

Kissling-Näf, 1999, pp. 23-26, modified by the authors

in the DMP of the FUG takes place. The model is situated in its socio-cultural environment, the institutions that characterise it, and the natural resource base that is the object of the DMP. The right side of the model shows the different steps between the formulation of decisions, their enforcement by the concerned people, and the eventual output. The arrows in the figure represent the dynamics of the system. They do not, however, indicate the kind, direction or importance of the interactions between the elements. It is the purpose of this paper to study these interactions.

Arena: actors and their situation

Gilmour and Fisher (1991, p.75) point out that '(c)ommunity forestry is concerned with control of resources, i.e. it is a political issue'. Starting from this point of view, it is interesting to introduce the concept of the 'policy arena'. Schmithüsen (1996, p.91) defines this concept in following way:

> The policy arena concept characterises the style and the manner of political processes which are linked with the settlement of conflicts and the ways of achieving a consensus.[3] (translation by the authors)

The arena represents the space where the actors elaborate their strategy and act in specific situations. The actors' positions, and the information to which the actors have access, are significant characteristics of these situations. The results of these interactions are the decisions that are taken within the institutional framework.

Two categories of actors related to the FUG are defined in this paper. The first one is that of the institutional actors, who are recognised and defined in the constitution of a FUG (e.g. the members, the assemblies, and the executive committee of the FUG). The second category consists of the organisations and people who are not mentioned in a constitution, but who want to express and defend their interests (e.g. NGO or community forestry projects like the Nepal Swiss Community Forestry Project — NSCFP). The District Forest Office (DFO) has an intermediate status (see fourth section).

3. 'Mit dem Begriff Politikarena werden die Art und der Ablauf politischer Prozesse gekennzeichnet, die mit der Austragung von Konflikten und der Konsensfindung verbunden sind'.

Institutions

The concept of institution used in this paper corresponds to that defined by Schmithüsen (1996, p. 58):

> The term institution [...] generally points towards persistent structures which are ordering relations between people, thus gaining an evolutionary advantage. At the individual level institutions lead to certain expectations in human behaviour. Due to their relative persistence they are in a position to provide people with a feeling of security with regard to common patterns of meaning and social action.[4] (translation by the authors)

An important point of Schmithüsen's definition is the emphasis on the long-term stability of institutions.

Natural resources

The concept of natural resources used in this paper corresponds to the definition of Kissling-Näf (1999, p.8):

> From a social-science perspective the term and concept of a 'resource' emphasises that nature or, the environment are only significant if they consider human needs.[5] (translation by the authors)

A resource is significant only if it satisfies somebody's needs (e.g. petroleum was not a resource in the eighteenth century). Since this study focuses on FUGs, the resources considered further are mainly the natural resources of the CF. A CF could be bush land as well as a dense stocking of big trees. The term 'forest' has to be considered in a broad sense. Other kinds of woodland like governmental forests or trees on private lands (agroforestry) have also been taken into

---

4. 'Der Begriff der Institutionen [...] verweist in der Regel auf überdauernde Einrichtungen in der zwischenmenschlichen Ordnung, die dadurch einen evolutiven Vorteil gewinnt. Institutionen führen auf der individuellen Ebene zu bestimmten Verhaltenserwartungen. Sie sind aufgrund ihrer relativen Konstanz in der Lage, ein gewisses Mass an Sicherheit bezüglich gemeinsamer Deutungs- und Handlungsmuster zu vermitteln'.
5. 'In einer sozialwissenschaftlichen Perspektive wird mit dem Begriff und Konzept der Ressource darauf hingewiesen, dass die Natur oder Umwelt nur bezogen auf menschliche Bedürfnisse eine Bedeutung hat'.

account, since they provide natural resources similar to the CF.

Management of natural resources

The management of natural resources is defined as the elaboration and organisation of measures related to the protection, conservation, development and exploitation of the CF, as well as the distribution of its benefits according to clearly formulated objectives.

Participation in the DMP

The notion of participation could apply to different levels of managing natural resources (i.e. DMP, enforcement of decision, monitoring). This study focuses mainly on the participation in the DMP and on the members of the FUG as actors in this participation. The definition of problems and the DM constitute the DMP.

Disadvantaged groups or people

The term *disadvantaged* denotes the target group of this study. H.L. Paudel (1998, p.7) uses the term 'less privileged people' (LPP), which he links with their economic and social status, as well as with their health. He asked the members of the three FUGs he studied to define LPP by themselves using PRA tools (Participatory Rural Appraisal). The members chose three criteria: land ownership, the number of cattle they own, and the school level. People belonging to the category of the less privileged were called '*nimchhara*' (Nep.). Ten staff members of the NSCFP had also been asked before the fieldwork for this study was started to come up with criteria that define disadvantaged people. PRA tools were also used and caste, land ownership, food sufficiency, school level, women's rights and income turned out to be the most significant indicators. Disadvantaged/advantaged and privileged/underprivileged are used below as synonyms.

Methodology

This case study analyses actors and institutions in their own context related to the management of natural resources. The selection of the study cases is the subject of the first section; the methods used to gather the information will be explained in the second; and the subject of the final section is the method employed to analyse the data.

SELECTION OF THE CASE STUDIES

The selection of case studies chose two districts where the NSCFP has its activities: the districts of Dolakha and Ramechhap.[6] Considering the limited time frame of 4 months with 5 to 6 days for the fieldwork per case study, the number of FUGs visited was limited to four. The selection criteria were:

- The FUG has to be located in the mid-hills, 'where there is extreme pressure on the forest and where livelihoods are most intimately associated with forests' (Hobley, 1996, p. 22).
- The CF had to be handed over before 1997. The members should have had the opportunity to gather experience in the participation in the DMP of their group for at least 2 years. FUGs formed before 1994 were not taken into account because '(t)he quality of the O(perational) P(lan) and constitutions formed at that time was quite poor' (NSCFP, 1999).
- The number of households of the FUG should not be too small (not less than 50), because the DMP is likely to be more informal, the smaller the group is. The group should not be too big (not more than 150), since only 10 to 15 people were interviewed. At least 10 per cent of the households had to be *untouchables* (Kami, Sarki and Damai), which was used as one criteria to define disadvantaged people.
- Politically sensitive areas were not to be taken into account.

The selection was based mainly on information provided by Lafranchi (1998) in the constitution and Operational Plans (OP) of the FUG. The data on the formation of the FUG was collected from the DFO of Dolakha and the District Co-ordination Office of the NSCFP in Ramechhap.

---

6. The third district where the NSCFP supports activities was not taken into account because of the relatively recent joint implementation of the PCF with the District Forest Office (DFO) and the poor knowledge the villagers concerning the projects (Dahal, 1999, pp. 26-27).

TECHNIQUES TO COLLECT INFORMATION

Primary sources of information

The main source of information was the structured individual interview. This technique was chosen to consider the DMP directly through the participants themselves rather than compare interviews of respondents from different FUGs. Besides the individual interviews, other techniques were used to consider different aspects of the problem, and were designed to enable people to participate actively (see Jackson and Ingles, 1995, for information about RRA and PRA). These techniques were participatory mapping to let the villagers visualise their CF, wealth-ranking to acknowledge the socio-economic situation of the FUG, visited and discussions concerning the question of access of disadvantaged people. The documents of the FUG (constitution and OP, minute book) were consulted, and semi-structured group interviews with three to five members of the EC were conducted for cross-checking the information.

Structured interviews

The structure of individual interviews used during the fieldwork was designed according to the following considerations[7]:

(a) The person and the FUG (the process of becoming a member of a FUG and investigations about the activities of the group).

(b) The person and the CF (the satisfaction of the needs and different functions of the forest since it became an officially recognised CF).

(c) Participation of the person in the DMP (the attendance of assemblies, voicing of interests, and influence on the decisions taken by the group).

(d) Participation in the DMP (from the point of view of the members of the executive committee, difficulties of the members of the FUG to take an active part in the DMP and possible improvements).

---

7. There was a broad diversity of languages involved in the conception and utilisation of the form, since nine people of different mother tongues were involved (French, German, English, Swiss Italian, Nepali, Limbu, Tamang, Magar, Newari). The answers were noted down in Nepali and translated into English afterwards.

(e) Perception of the forest (description of the forest).

(f) Socio-cultural attributes (caste/ethnic group, age, school level, economic status, and experiences in an executive committee).

A PRA exercise (matrix ranking) was used to assess the degree of satisfaction with forest products (point b). The selection of the villagers to be interviewed was done considering the first main research question (i.e. participation in the DMP, with particular respect to the disadvantaged people). It was not possible to work with statistically representative groups, due to the great heterogeneity which characterises the FUG, the poor information that was available, and the limited time. The disadvantaged people and the members of the executive committee as privileged actors of the DMP were two target groups for the interviews. A third group takes the other members of the FUG into account to compare them with the answers of the two other groups. The criteria of the selection were based mainly on caste and ethnic identity:

- At least 10 people per FUG were to be interviewed.
- Three untouchables members of the Damai (tailor), Kami (black-smith) and Sarki (leather-worker).
- At least three members of the executive committee, of which one was a woman or an untouchable.
- Two members of the prominent castes in the village, of which one is considered as disadvantaged (*nimchhara*) by at least three people of the FUG.
- Members belonging to other ethnic groups or castes.
- The selection of interview partners within the families was done by considering those who take the decisions in the FUG on behalf of the family.

DATA ANALYSIS

This section presents the methods used to analyse the answers from the individual interviews. The information gained from other techniques has been analysed qualitatively. The answers of the interviews have been structured and compiled in a database (Microsoft Access 7.0).

Socio-cultural status

The analysis of socio-cultural status aims to determine three categories, which should define groups of people related to their status (disadvantaged, intermediate and advantaged). These categories are of great importance, since they represent the key to analyse the answers of the other questions (points a - e) by creating pivot tables. The affiliation of an interviewed person to one of the three categories is based on different socio-cultural attributes that had to be defined after having considered the results of the wealth-ranking exercises. The interviewed people's attributes are put into three categories: low, intermediate and high. 'Low' corresponds to a value below which the 9 to 11 people with the lowest values of the attribute are located (e.g. if the 9 poorest interviewed people have less than 5 ropani, the criterion will be this value). The criterion to determine the attribute 'high' is defined in the same way. The flexibility to choose between 9 and 11 people, a quarter to a third of all interviewed people, allows one to define more or less homogeneous groups (e.g. if nine people have less than five cattle, and the two next one have 10 cattle each, the criterion 'low' will be the ownership of not more than five cattle). Finally, the attribution of an interviewed person to a status category is based on a ranking system in which each attribute gets a number of points, which are positive if the person belongs to the category 'high', or negative in the opposite case. The sum of all the points defines the person's status.

Results and analysis

PRESENTATION OF THREE CASE STUDIES[8]

All FUG which were investigated are from the Taamaa Koshi valley. The settlements are located mainly between 900 and 1600 m above sea level. The FUG of Bhudha Bhimsen is the only one which has a direct road connection with the Lamosangu-Jiri road. The FUG is about half an hour away from Charikot, the district headquarters of Dolakha, and can be reached in about 8 hours by bus from Kathmandu. It is necessary to walk at least half an hour from the other FUG to reach the nearest road and then to travel about one and half hours to reach Charikot.

8. Sources: Operational Plan, wealth ranking, Lafranchi (1998).

About 130 households are members of Budikhoriya FUG, 160 of Bhudha Bhimsen FUG, and 200 of Piple FUG.[9] Two-thirds of the households of Budikhoriya FUG are Shrestha (Newar ethnic group) while two-thirds of the households of Bhudha Bhimsen FUG are Chhetri. The Chhetri represent the main caste in Piple, making up about one-third of the population. The Piple FUG has the greatest diversity of castes and ethnic groups — not less than 10 — of the three visited villages (see *fig*. 2).

There are very few livelihood opportunities besides agriculture on terraces. Emigration is significant in the three FUGs[10] and about

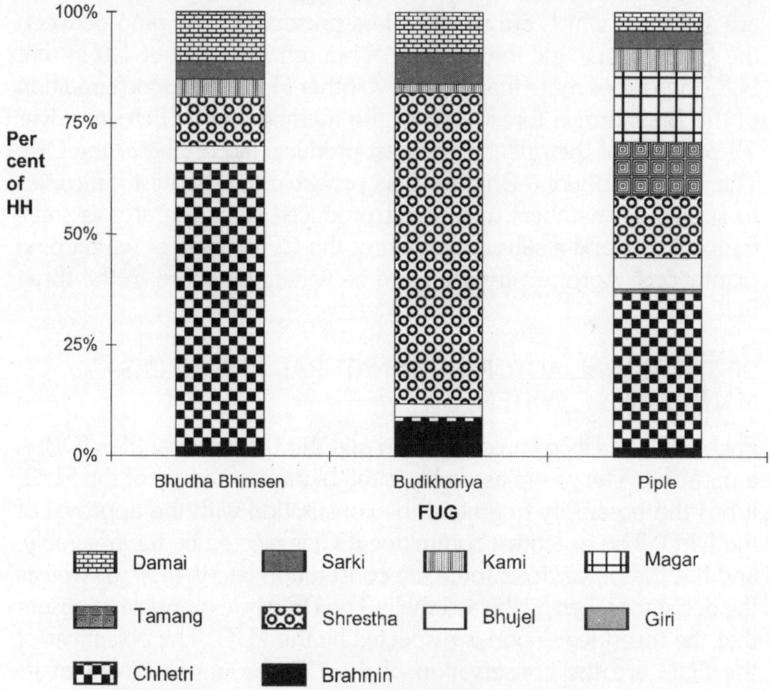

***Fig.* 2. Ethnic and caste composition of the three FUG.**

9. Seventy-five households became members of Piple FUG in 1998. These people are now living in another VDC, one hours walk away from the centre of the FUG, and have only few contacts with the other members. They were not considered in the study.

10. Working from the hypothesis that there are not more than two people per household who emigrate, the proportion of families concerned by this phenomenon is between 20 and 25 per cent of all households in the FUG.

40 to 70 people per FUG work outside their village. About 60 per cent of the members of the FUG of Bhudha Bhimsen and Budikhoriya and 45 per cent in Piple own less than 10 ropani of land (20 ropani =1 hectare). This criterion has to be considered carefully given the fluctuation of the productivity of land within a FUG (depending for example on the altitude), and between the three FUG. Only 10 per cent of the people in Budikhoriya had a school leaving certificate (SLC) in 1995. This rate was lower in the two other FUG. All three community forests (CF) had been overexploited before the formation of the FUG. The CF of Piple is the biggest one with 200 ha and although it is degraded in the surroundings of the habitations, there are still parts which are more or less preserved. The ratio between the CF area (ha) and the number of households is about 1.0 in that FUG, two times more than in the two other FUG. The good condition of the Budikhoriya forest enables the members of its FUG to cover 75 per cent of their needs by forest products (according to the OP). The FUG of Bhudha Bhimsen has probably the greatest difficulties to supply its members with forest products, since the area is small (ratio of 0.5) and a substantial part of the CF consists of young pine plantations. Agroforestry seems to be widely practiced in the three FUG.

INSTITUTIONAL ACTORS AND NATURAL RESOURCES MANAGEMENT SYSTEM[11]

Each FUG has its own constitution and the Operational Plan (OP) is a part of it. The group assembly is the highest authority of the FUG. It has the possibility to amend the constitution with the approval of the DFO. The executive committee (EC) is elected by the assembly, and has the power to enforce the constitution and the OP, as well as the decisions taken by the assembly. The DFO is responsible to ensure that the forest legislation is respected by the FUG. The objectives of the FUG are the conservation of the CF, the improvement of its conditions, and its use and management for the benefit of the members.

The EC disposes of an OP to manage the CF, which is according to the FUG more or less detailed (in decreasing order: Budikhoriya, Bhudha Bhimsen and then Piple). The EC proposes the

11. Sources: FUG constitution, Operational Plan, semi-structured group interview.

implementation of decisions taken by the assembly. The users themselves serve as watchmen for their CF in a rotation system (Budikhoriya and Piple), or alternatively a watchman is paid by the FUG (Bhudha Bhimsen). A sanction system has been set up to fine people breaking the FUG rules. The users have to help in fighting forest fires, as well as in measures to stabilise erosion-prone slopes. Plantations on uncovered land or enrichment plantings (fruit trees, fodder trees) are also carried out to improve the forest condition. Silvicultural operations are accomplished by all members[12] who are remunerated by forest products for their work (nothing is lost in a society depending on wood as the main source of energy), and timber is distributed considering individual demands.

DYNAMICS OF THE THREE FUG[13]

The EC of all three FUG gets together 8 to 12 times a year to take decisions, with an average of three per meeting. About the same number of decisions are taken during the assembly sessions of Piple and Budikhoriya, which is not very frequent in comparison to Bhudha

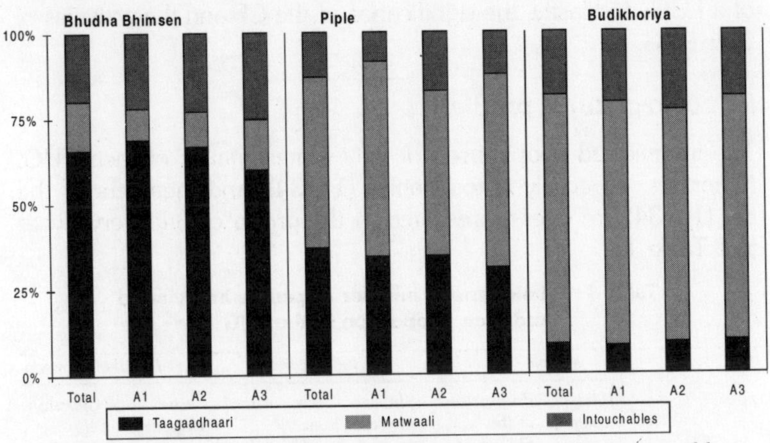

***Fig.* 3. Attendance of ethnic groups and castes at three assembly meetings**

Total: corresponds to the proportion of the cultural group (*taagaadhaari*, *maatwaali*, untouchables) in a FUG independently of the attendance of its members at an assembly (see *fig*. 3). A1: last assembly; A2: second last assembly, etc.

12. Except in Piple: the users have to pay a tax of Rs. 5 per day for work not accomplished.
13. Sources: minute books of the last three years.

Bhimsen (where 16 decisions are taken on average). The frequencies in which assembly sessions are held could partly explain this difference (one ordinary assembly a year in Bhudha Bhimsen in comparison to two or three in the two other FUG). The decisions taken are not limited to the CF management, but also deal with the development of the community. Some decisions were taken in Bhudha Bhimsen to regulate conflicts that had nothing to do with the forest. The rate of attendance at the assemblies is the lowest in Piple (about 45 per cent) and highest in Bhudha Bhimsen and Budikhoriya (about 75 per cent). The quorum in the constitution could explain the high proportion (about 2/3) in the two other FUG. There is no cultural group that is under or over-represented in the assembly meetings (see Table 1).

THE PARTICIPANTS

This section focuses on the participants in the DMP (i.e. the users and the members of the EC).[14] The socio-cultural profile of interviewees is first examined; then the process to become a member of a FUG; and lastly, the significance of the CF and the activities of their FUG.

(a) Socio-cultural profile

The interviewed people are not fully representative of their FUG. Minorities, especially untouchables (11/34), and members of the EC (10/34) are over-represented in the group of 34 interviewees (see Table 1).

Table 1: Ethnic groups number of people interviewed and their proportion in the FUG

	Piple		*Budikhoriya*		*Bhudha Bhimsen*	
	Inter-viewees	*Proportion in the FUG (HH)*	*Inter-viewees*	*Proportion in the FUG (HH)*	*Inter-viewees*	*Proportion in the FUG (HH)*
Taagaadhaari	3 (25%)	37 %	1 (10%)	9 %	5 (42%)	66 %
Maatwaali	6 (50%)	50 %	5 (50%)	73 %	3 (25%)	14 %
Untouchables	3 (25%)	13 %	4 (40%)	18 %	4 (33%)	20 %
Total	12(100%)	100 %	10 (100%)	100 %	12 (100%)	100 %

The attributes to define disadvantaged people during the 'wealth ranking' exercises were mostly linked with economic status: land

14. Sources: individual interviews.

ownership, number of cattle, food sufficiency, and income opportunities. The school level was taken into account, too. These attributes correspond to what Paudel and the interviewed members of NSCFP have mentioned. Considering this, the attributes and criteria used to define to which social category the people belong are given in Table 2. The experiences in the DMP gained by people involved in an executive committee have also been taken into account. The parameter considered is the membership in an executive committee and the position the person has (e.g. chairperson, vice-chairperson, secretary, treasurer). The ethnic group has not been taken into account in the definition of the social status.

Table 2: Criteria to define the social status of the 34 interviewed people

Attributes	*Points*	*Criteria*	
		Disadvantaged	*Advantaged*
Land ownership	½	Less than 5 ropani	More than 20 ropani
Ratio 'rop/PP'*	½	< 0.5	> 2.0
Number of Cattle	1	< 2	>= 5
Main activity	1	Labour[15]	Regular income
School level	1	Illiterate	SLC
Member of an EC**	½-1½		EC member (½) with a position (1)
The total points determine the social status ***:		Disadvantaged: less than -2.5	Advantaged: more than +2.5

* Not only does the land ownership seem to be an important attribute to define social status, but also the total number of ropani divided by the number of people depending on it. For example, somebody answered that he owns 62 ropani of land and explained that 23 people depend on it. Therefore this attribute has been split into two: Land owning and Ratio 'Rop/PP' (Rop: ropani; PP: people depending on this amount of ropani).

** A person could be a member of different executive committees. The maximum point score 1½ .

*** Points obtained in the disadvantaged category are negative. They are positive in the advantaged category. All points are added and the total determines the social status of a person.

The main outputs of the analysis of the socio-cultural attributes related to the FUG, the cultural groups, and the social status are now summarised.

15. People who are working on request.

Forest user groups

- The people interviewed in Bhudha Bhimsen are mostly able to read and write Nepali, but did not attend a school. Half of the people interviewed in Piple have been to school and, with one exception, the others are illiterate.
- The FUG of Budikhoriya is characterised by the highest proportion of poor and disadvantaged people (5/10), while Piple has the lowest ratio (1/12).

Ethnic groups

- No member of a *taagaadhaari* caste (the high-ranking so-called 'twice born' castes such as Chhetri, Brahmana) is illiterate, poor or disadvantaged, whereas no untouchables went to school.
- Most of the *taagaadhaari* (6/8) were, or are, members of an executive committee.
- The *maatwaali* (the 'alcohol drinking' ethnic groups such as Shrestha, Tamang, Magar) are an intermediate group, since no particular feature could be recognised. Most of the *maatwaali* are Shrestha (10/14).
- The ethnic group with the most interviewed people are the Shrestha (10). Among them there are advantaged as well as disadvantaged.

Social status

- Nobody who is untouchable, illiterate, poor or without experiences of an executive committee membership, is considered as advantaged (see *fig.* 4).
- Nobody who has a SLC, is a member of a *taagaadhaari* caste, rich or member of an executive committee, is considered as disadvantaged.
- The majority of the interviewed people who are disadvantaged (5/9) was found in the Budikhoriya FUG.

Disadvantaged people have very few cattle and land in comparison to the others (see circle in *fig.* 4).

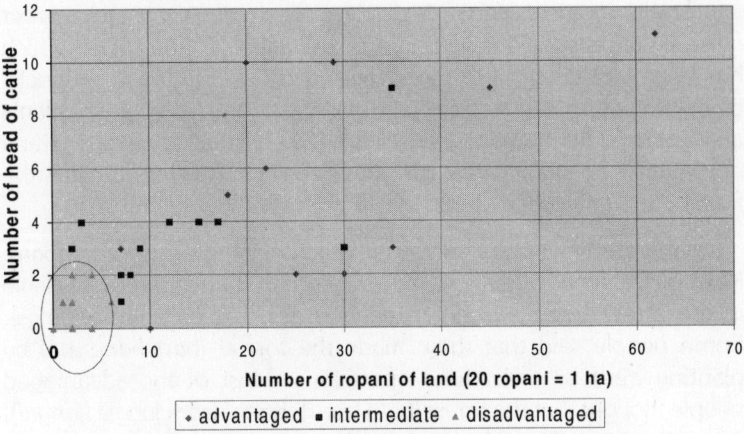

***Fig.* 4. Social status and economic parameters (cattle and land ownership)**

(b) Perception of the forest

The perception of the forest and forest-related phenomena was studied by asking people to explain terms it (forest, jungle, tree, shrub). They were also asked whether they went alone into the forest, since informal discussions prior to the fieldwork showed that people were sometimes afraid to go there alone.

The use of structured interviews has shown its limitations regarding the perception of forests. The subject was obviously new and difficult to speak about for the interviewees. They were asked to explain a part of their environment that they are very used to seeing and interacting with. Interestingly, the school level seems to correlate to the number and diversity of the answers (see Table 3).

Table 3: School level and diversity of the answers

School level	*Number of elements mentioned on average*
School leaving certificate	3.2
Have been in school	2.7
Literate	1.8
Illiterate	0.9
All levels	2.0

The number of elements denotes the frequency of an answer given by an interviewee (during an explanation of the forest) which has been sorted out in the following categories: (i) forest products; (ii) infrastructure and services (e.g. erosion, water supply); (iii) plants and animals; (iv) human intervention (i.e. plantation or silvicultural operations); (v) significance for human beings; (vi) 'the forest is the forest'; and (vii) others.

Different ways were used to characterise the term forest (*ban*). Most of the time it was done by mentioning the resources available in the forest and by its constituents (trees, shrubs and seedlings). Some people said that they 'made the forest' (*ban banaune*) by planting trees or silvicultural operations. Most of the advantaged people thought that the forest (*ban*) was different from jungle (*jangal*), in contrast to other interviewees. There was a general anxiety to enter the 'jungle' due to dangerous wild animals. The majority of the interviewees in Piple did not enter the forest alone, probably owing to its large size (200 ha).

(c) Access to a FUG

This section aims to study the motivations and the difficulties people have to become a member of a FUG and the changes in their perception of the CF after they have become a member. All interviewees became members of their FUG at the time of its formation and no one complained of difficulties.[16] One-third of the people said that they discussed the issue with their family before deciding to join. The majority of the untouchables did not discuss it with anyone. The main perceived reasons to join a FUG were to get forest products and to protect the forest. While all the advantaged people mentioned forest protection, only few disadvantaged did. The same could be observed between *taagaadhaari* caste members and untouchables (see Table 4).

16. This does not mean that nobody experienced difficulties. While talking informally in tea-shops, it appeared that membership to a political party or to an administrative unit other than the one were most of the FUG members live, or the emigration of a husband, prejudiced the likelihood of becoming a member of a FUG.

Table 4: Forest products and protection of the CF as reasons to become a member of a FUG

Social status	*Forest products*		*Protection*	
	Mentioned	*Not mentioned*	*Mentioned*	*Not mentioned*
Advantaged	8	1	9	0
Intermediate	7	3	4	6
Disadvantaged	8	1	2	7

More than 80 per cent of the people interviewed perceive recently-created community forests differently to the past. The reasons were the better forest condition, its protection against overuse, and easier access to forest products.

(d) Significance of the community forest

This section deals with the religious significance of the CF, as well as its significance to harvest forest products. Religious ceremonies are performed in the CF of Piple and Bhudha Bhimsen, but not in Budikhoriya. Most of the interviewees, however, do not seem to go to the CF to perform rituals.

Firewood is considered by far the most important forest product besides timber and fodder, because it is the main energy source. Kerosene is rarely used and electricity, if available, is for lighting only. Few chemical fertilisers are used in Piple and Budikhoriya, but in Bhudha Bhimsen people reported that they degraded the quality of the soil.

CF and trees in private fields are the main sources of supply of forest products. Advantaged people mostly considered their private fields their main source of subsistence. The management of the CF seems to improve the quality of the forests, reduce over-exploitation, and most of the people think they will get more forest products in the future.

(e) Satisfied needs and consented investments

The aim of this section is to study people's satisfaction considering the supply of forest products they get from the CF, and the investment in terms of money and time they have to provide. Two-thirds of the interviewees wanted more forest products, especially firewood, timber and fodder. The main reasons given were their limited availability on private land. In particular, timber to build or maintain houses was in high demand. Most of the disadvantaged people said they needed more forest products. Among the three FUGs, the highest demand for extra forest products was found in Bhudha Bhimsen, probably because it had the smallest ratio of land per household, and only had young pine plantations in its CF.

The activities normally performed within the FUG are: guarding the CF, silvicultural operations, and harvesting of forest products. 80 per cent of the interviewed people do not think they have to work too much in the CF. The reasons mentioned were the participation of all users in these activities, the possibility of getting forest products, and the improvement in the forest condition. Half of the respondents thought that the prices for forest products were not high. Most people who disagreed with the prices were in Budikhoriya (6/10) and the fewest were found in Piple (2/12). In contrast to the advantaged people (4/11), disadvantaged people (5/8) answered that the prices were too high.

(f) Protection of the CF against misuse

Ostrom (1990) suggests that the protection of a CF against misuses is an important factor in the stability of an institution. If the users feel that the CF is not protected well enough, they will probably not be interested in spending more time and money on it without being sure of getting adequate benefits. The majority of the people think their CF is sufficiently protected against the misuse by outsiders (70 per cent) or the misuse by insiders (95 per cent). Problems with the identification boundaries of the FUG may explain the first percentage concerning transgressions by outsiders. According to Mr. Sitaram Basnet, Chairperson of FECOFUN Dolakha,[17] this problem is probably the most important the FUG have to face at the moment.

(g) Knowledge about the FUG fund and opinions about its utilisation

The utilisation of the FUG fund was a subject of intense discussions since people were given the opportunity to discuss expenditures (e.g. for the construction of a temple, support of poor families) and income from penalties.

17. The Federation of Community Forest Users in Nepal (FECOFUN) is the nodal NGO related to FUG.

The three FUG constitutions contain provisions related to the establishment of a FUG fund and its management. A bank account has to be opened and a treasurer appointed who is responsible for its management. At least once a year, the fund accounts have to be presented at an assembly session and a copy sent to the DFO. The users have the right to refuse them and to entrust a third person with an audit. The main incomes come from the sale of forest products, penalties, loans and subsidies from the government or organisations. At least 25 per cent of the expenditure has to be invested to develop the CF according to the Forest Act, 1999.

The amounts of the funds of Bhudha Bhimsen, Budikhoriya and Piple were respectively, NRs 60,000, 83,000 and 85,000 which is by far from negligible. A labourer earns about NRs 50 to 70 per day. While at least two-thirds of the advantaged people knew the amount of money in their fund,[18] the way it was used, and the name of the responsible person; not more than one-third of the disadvantaged people were able to answer these questions correctly. Considering their socio-cultural attributes, untouchables, illiterate and poor people answered like the other disadvantaged people. Nobody answered that he/she did not agree with the way the fund is used, perhaps because of the formulation of the question.

PARTICIPATION

There are no provisions in the three constitutions related to the way collective decisions have to be taken.[19] In fact, the EC decides on the agenda of assembly meetings and presents it at the beginning. Users have the possibility, however, of complementing it with their own requests. Decisions are taken by consensus or by majority. Almost all users think that the assembly meeting is the right venue for decisions to be taken. Three levels of participation in the DMP have been considered further:

- attendance of assembly meetings — participating in the DMP.
- definition of problems — putting forward interests.
- taking decisions — influencing the DMP.

---

18. With a precision of +/- 10 per cent.
19. Sources: individual interviews, semi-structured interview and FUG constitution.

The results presented in this section reflect the outcome from individual interviews. Due to their privileged position in the DMP, EC members have been asked other questions than the users. The main focus of the questions was on difficulties the users might experience while taking part in the DMP. However, the possibility of considering participation from a different angle leads to the formation of two groups: the user group with 24 people and that of the EC members with 10 people. The pertinence of the results decreases with the decreasing number of people per group. Eight advantaged people out of twelve interviewed were EC members of their FUG at the time this study was conducted.

The number of users who attended the two last assembly meetings was lower in Budikhoriya than in the other two FUGs. No particular trend in participation was noted concerning the ethnic groups. The main reason not to attend an assembly meeting was other urgent work that had to be done. Only half of the users, of which most of the people are from Bhudha Bhimsen, remember an important decision taken during the last two assembly meetings. The majority of the users did not come forward with a request on the occasion of the two last meetings. The main reason could be related to the *thulo maanche* pheno-menon. *Thulo maanche* literally means 'big persons' and defines people who have an important position in the society, whereas *saano maanche* ('small persons') is used for ordinary people.

Budikhoriya FUG, where all decisions of the three last assemblies were taken by the EC, is characterised by the largest proportion of users who expressed a request. In Bhudha Bhimsen it is just the other way round: only one-third of the users felt that they were able to influence the decisions taken within their FUG, and none of the disadvantaged expressed this feeling. The reasons are probably also related to the *thulo maanche* phenomenon. It is not appropriate for a *saano maanche* to take an active part in the DMP from the point of view of a *thulo maanche*.

Synthesis of the results and discussion

This last section aims to answer the two research questions, concerning the perception of the forest and the factors that have an influence on the DMP. The results of this study are not representative at the overall group level (only a few people have been interviewed), nor at the district level (only 3 FUG have been studied in two districts). However, the results point to trends and factors that are likely to be significant for participation in the DMP.

PERCEPTION OF THE FOREST

The religious dimension of the forest seems to change, especially in Budikhoriya where the people went into the CF area to perform religious activities in the past, but no longer do so. CF and agroforestry, apparently well practiced in the three FUG, are sources for the supply of forest products. Rich people seem to be less dependent on the CF. Firewood, timber and fodder are the forest products the interviewees think are the most significant, and seem to have no important substitutes. Most of the people require large quantities of these products, especially the people of Bhudha Bhimsen FUG, where the demand is the highest (11/12), but has the lowest potential to fulfil the users' needs. Paradoxically, potential is probably the highest in Piple, where the demand is the lowest. Most of the respondents in Piple felt they had enough products, yet the lowest proportion of adherents to their FUG was also found in Piple. There is thus a logical relation between the level of the fulfilment of the users' needs in forest products and the conditions of the forest.

People's perception of the forest was by far the most difficult to speak about. Different explanations of the term 'forest' were given: first through the notion of a 'resource', and then through its constituents. It is not surprising that the forest was described as a resource. For example, 80 per cent of the people mentioned the possibility of getting more forest products as a reason to adhere to a FUG. Some people explained the term 'forest' by describing how they planted trees or managed silvicultural operations. Some people saw 'forest' as a space mastered by man, while 'jungle' was seen as a wild space which is not only dangerous to enter, but sometimes also impenetrable. This distinction between a managed forest and a wild jungle could be studied further.

FACTORS THAT INFLUENCE PARTICIPATION IN THE DMP

The assembly has a privileged position in the political arena, since important decisions of a FUG are taken within its structure. The assembly session is an occasion for all users to meet, but the DMP is not limited simply to this institution. The executive committee, the second institution of the FUG, enforces the decisions taken by the assembly and accomplishes its duties (e.g. sanctions forest offences, or distributes forest products on individual demand). Most of the questions about participation in the DMP were formulated without explicitly mentioning the assembly or the executive committee, in order not only to study the formal part of this process, but the informal as well. The DMP and the factors which have an influence on the participation of the users is now discussed.

(a) Informal discussions

Public places like tea-shops or agricultural fields offer the forest users the possibility of communicating about the FUG activities. There are three tea-shops in Bhudha Bhimsen, and one, open in the morning, in each of the two other villages. Two people first mentioned their demands regarding decisions taken on forest matters in such locations before voicing them in an assembly meeting. During informal discussions, sensitive matters can be more easily addressed. These meeting places seem to encourage the formulation of problems and opinion-making. Some users said that they had first approached an important person (*thulo maanche*) to get his support in the formal meeting.

(b) The executive committee

All users have the opportunity to attend the meetings of the executive committee. The reasons to do so are manifold, such as to be notified of a sanction, or simply to listen to the deliberations. On these occasions, the users can submit a request to the EC which decides whether or not to include it in the agenda of an assembly meeting.

(c) The assembly

Each user, or more precisely, each representative of a household, has the right to attend the assembly meetings to either: (i) submit a request; (ii) take an active part in the deliberations and the decision-making; or (iii) vote on the FUG constitution.

(i) Attendance at assembly meetings

The attendance at the assembly meetings is the highest in Bhudha Bhimsen and in Budikhoriya where about 75 per cent of the households are represented, whereas it is only 45 per cent in Piple. The lack of a quorum in the constitution of Piple could explain this difference. The analysis of the ethnic and caste composition of the people who attended the three recent assembly meetings show that no ethnic group is under- or over-represented on these occasions. The reasons mentioned for not attending an assembly were in most cases lack of time due to an intensive workload. Some of the male household-heads who did not take part in an assembly meeting asked their wife or one of their elder children to go. Statistics about the attendance rate show that 20 to 40 per cent of the participants at an assembly meeting are women. The participation of women in the DMP has almost not been taken into account in this study (only one woman was interviewed). For further research, it would be useful to investigate women's role in the decision-making process of their FUG.

(ii) Definition of problems and issues

All interviewees who submitted a request to the FUG (9) have done so in an assembly meeting. All the decisions taken in Budikhoriya were submitted by the EC. This was also the case in Piple except for one decision concerning a request to build a temple on the proposal of a local organisation. About one-third of the decisions taken in Bhudha Bhimsen had been suggested by users. On average, four decisions have been taken in Budikhoriya and Piple during the last three assembly meetings. This average amounts to 16 in Bhudha Bhimsen. The high attendance at the assembly meetings, the many informal meetings, and the high demand for forest products that characterise Bhudha Bhimsen FUG would explain why more decisions are taken during its assembly meetings, and that the users seem to be more active in the DMP. Among the 24 interviewees, about two-thirds answered that they had not come forward with a request. The main reason mentioned was that they thought that it was useless to do so and some users complained that important people would not listen to them.

(iii) Decision-making

Less than one-third of the respondents (7/24) answered that they were able to influence the decision-making in their FUG. No disadvantaged people belonged to this category. The reasons given were linked to the *thulo maanche* phenomenon of deeply-entrenched social roles ascribed to in a Nepalese village community. It is also interesting to look at this phenomenon from the point of view of the 10 interviewed EC members. Six respondents denied that users have such difficulties. It seems that there is a lack of communication between the EC members and the users about this problem. Another problem seems to be the lack of interest of the users to take an active part in the DMP. There was no question about this problem in the structured questionnaire, but two people mentioned it during the individual interviews. One EC member of Budikhoriya and one forest user of Bhudha Bhimsen complained about the fact that users do not pay attention during an assembly meeting. The percentage of people (20-40 per cent)[20] who answered that they do not remember an important decision taken during the last two meetings is remarkable. The assembly meeting is also a social event and could be a reason why users do not pay attention to the debates. Assembly meetings offer a possibility to meet each other and speak about matters other than that of the CF.

The purpose of the debates held during the assemblies are perhaps a key to interpret this lack of attention. The users have to be motivated to break into the politic arena which the assembly meeting is a critical part of. Such a motivation could be in the form of a new rule, a new decision that affects them either negatively (e.g. rising taxes) or positively (e.g. temple construction), as well as the change of a rule that favours their social position. The aims of a FUG are likely to be linked to the management of the CF whose targets are first of all the conservation and utilisation of the community forest. These two targets require an optimum balance between exploitation and complete protection. Yet, this optimum is a political issue and has to be negotiated in the political arena of the FUG in connection with matters affecting the whole village community.

The motivation to become a member of a FUG was mainly to get access to forest products and, to some extent, to protect the forest. The correlation of the social status of the respondents with their motivation to manage the CF shows an interesting difference

---

20. Five people answered that no important decision had been taken during the two last assemblies, and five other answered they did not know (24 people were asked).

between actor groups. Protection is the principle motivation for advantaged people (9/12), while only a few disadvantaged people mention it (2/9). Most of the advantaged people also feel that the capacity of the forest is not sufficient to cover their needs for forest products. The fact that almost all advantaged people (11/12) were or had been members of an EC could explain their higher concern for the CF protection. As an EC member, they probably felt more responsible for the protection and conservation of the CF which is a legal condition to the right to manage their forest (the government has not handed over the property right on a permanent basis).[21] Because of their formal position (EC member) and informal status (*thulo maanche*), they are likely to impose their desire to protect the CF.

The access to forest products was the main motivation to join a FUG. Harvesting, the distribution of firewood and timber are major issues, especially for the disadvantaged people who strongly rely on them. However, the harvesting and distribution patterns are clearly defined and are not easy to change. Firewood is collected during silvicultural operations in which every user can take part and timber can only be harvested on individual demand. Even if the quantities from the CF are not enough to fully satisfy the needs of the group members, the users must accept it either because of the limited capacities of the CF, or due to the fact that all users have equal entitlements.

An other major issue could be the utilisation of the FUG fund, favouring the disadvantaged FUG members. Although no one openly disagreed with the current utilisation of the fund, the disadvantaged people were hardly aware of its existence. For instance, the disadvantaged people did not know the name of the treasurer, the amount in the fund, or the way it was used. The poorer people could profit more from the fund, by asking for a reduction in the fees they have to pay for forest products.

---

21. 'During (the) Fiscal Year [1999] an important political decision for the Community Forestry Program has been the amendment of the Forest Act (Forest Act, 1999). . . . (T)his Act allows the District Forest Office to take action directly against the FUG or member of the FUG and punish them if they have done something in transgression to the constitution and the Operational Plan of the FUG'. NSCFP, 1999, p. 4.

Conclusion

The participation in the DMP allows forest users to express themselves and to reflect the diversity of their interests in CF management. Participation also allows the users to influence the decisions taken by the FUG. The constitutional set-up for the participation of all users in the DMP seems to be known to all interviewed people, although some users, especially disadvantaged people, still encounter social and psychological barriers to make use of it. The difficulties are probably linked to a lack of basic formal knowledge (illiteracy); little awareness of their constitutional rights and duties; a lack of knowledge about the subjects that are debated (e.g. funds, management of the forest); and a lack of experiences with participation in a DMP. The fact that the recent democratisation process in Nepal has taken place mainly, if not only, at the national level, could explain this. Many users are probably still used to decisions that are taken by either government officials or important people, and that their own opinion, the opinion of the *saano maanche*, is not necessarily taken into account. Users should therefore not only get the opportunity to obtain the necessary knowledge and skills to take an active part in the DMP, but should also make use of the inherent political potential of the Nepalese community forestry programme. Whether this option can ever work in a hierarchical society based on caste and ritual purity remains to be seen.

Recommendations

The following recommendations focus on particular issues that should be studied further for a better understanding of the participation process:

- The change in perception of the forest induced by the formation of a FUG, especially by the elaboration of an Operational Plan.
- Knowledge the users have about forest management.
- The rights and duties involved in their FUG.
- The way the CF is managed by silvicultural operations and planning.
- User's perception of their social role within their FUG (c.f. *thulo maanche*), particularly with respect to women.

- Patterns of participation, especially with regard to finding consensus and its implications.

References

Dahal, G. R., 1999, *Implementation of the community forestry programme in the mid-hills of Nepal — Assessment of the present situation in Okhaldhunga district*, SDC/NSCFP.

Gilmour, D.A./ Fisher, R.J. 1991, *Villagers, Forest and Foresters: The Philosophy, Process and Practice of Community Forestry in Nepal*, Kathmandu, Sahayogi Press.

Graner, E., 1997, *The Political Ecology of Community Forestry in Nepal*, Verlag für Entwicklungspolitik, Saarbrücken.

Kryger, K., Mortensen, T.F., 1998, *Communiy Forestry in Nepal — A study of under-utilisation*, Roskilde University, Denmark.

His Majesty's Government of Nepal (HMGN), 1988, *Master plan for the forestry sector — Nepal*, Ministry of Forest and Soil Conservation, Kathmandu.

HMGN, 1993, *Forest Act 2049*, English translation.

HMGN, 1995, *Forest Regulations 2051*, English translation.

HMGN, 1998, *Statistical Pocket Book — Nepal, 1998*, HMG, Central Bureau of Statistics, Kathmandu.

Hobley, M., 1996, *Participatory Forestry: The Process of Change in India and Nepal*, Overseas Development Institute, London.

Jackson, B., Ingles, A., 1995, *Participatory Techniques For Community Forestry: A Field Manual*, Nepal Australia Community Forestry Project, Canberra.

Kissling-Näf, I., 1999, Ressourcenökonomie III, second draft, unpublished script, Professur Forstpolitik und Forstökonomie, ETH Zürich.

Lafranchi, S., 1998, The Economic Potential for Forest User Groups by Sustainable Production, Transformation and Commercialisation of some Non Timber Forest Products in the Hill Districts of Dolakha and Ramechhap, Nepal, unpubl. diploma thesis, ETH Zürich.

NSCFP, 1997, *Summary Report of General Assemblies of Community Forestry User Groups — Dolakha and Ramechhap*, NSCFP, Kathmandu.

NSCFP, 1998, *Nepal Swiss Community Forestry Project — Annual Report 1997/98*, NSCFP, Kathmandu.

NSCFP, 1999, *Nepal Swiss Community Forestry Project — Annual Report 1998/99*, NSCFP, Kathmandu.

NSCFP, 1999, *FUG Assembly Monitoring Result*, NSCFP, Kathmandu.

Ostrom, E., 1990, *Governing the Commons — The Evolution of Institutions for Collective Action*, Cambridge University Press, New York.

Paudel, D., 1999, *Distributional Impacts of Community Forestry Programmes on Different Social Groups of People in the Mid-Hills of Nepal*, University of Cambridge, Department of Geography, UK.

Paudel, H. L., 1998, *The impact of community forestry programmes on the life of less privileged people — three FUGs in Dolakha district*, Central Department of Sociology/Anthropology, Tribhuvan University, Kathmandu.

Schmithüsen, F., 1996, Gesellschaftliche Regelungsprozesse der Erhaltung und nachhaltigen Nutzung des Waldes — Unterlagen zum Fachgebiet Forstpolitik, Professur Forstpolitik und Forstökonomie, unpublished script, ETH Zürich.

Seeland, K., 1996, 'Sociological remarks on "community forestry" in Nepal'. In: Stellrecht, I. (ed.), 1996, *Contributions to Comparative High Mountain Research: Pakistan-India-Nepal*, Cultural Area Karakorum Scientific Studies No. 3. Verlag Koppe, Köln.

Tembe, D.K., 1998, *Forest Users' Groups : Social Foundation For Participatory Development in Mid-Hills of Nepal*, School For International Training, Brattleboro, Vermont, USA.

14

Environmental Perception and Forest Management among the Tamang of Southern Kavre

Buddhi Tamang
Amrit Yonjan

Background

NEPAL is known as one of the countries which has a mosaic of multi-ethnic and multi-linguistic societies. Amongst these, Tamangs are considered an indigenous ethnic group in the surrounding areas of the Kathmandu valley and the central development region of Nepal. According to Tamang (1994) and Bhattachan (1993), indigenous people comprise of those groups who continue to practice their traditional culture, religion and language even after having been colonised by outsiders. One of the main criteria in determining the nature of their cultural identity is their ancestral link to the land and water resources of a particular area. The Tamang population has a long history that is associated with the Tibeto-Burman speaking and Mongoloid populations. In this way, the Tamang form an integral part of the cultural mosaic of numerous ethnic groups residing in Nepal which have their own cultural, socio-economic, environmental and historical background. In the past, academic research on Tamang culture was mostly done by non-Tamang. The process of cultural and political self-examination and indigenous representation of their culture has begun only recently. This feeling has become more relevant and essential in the changing democratic scene of the recent political changes in Nepal, and is apt to secure their inherent rights and bring them into the national mainstream.

There is a need for a careful assessment of the external factors influencing Tamang socio-cultural development that are guided by state policies. One important aspect is the internal preservation process of the Tamang: origin myths, social rituals, oral history, and their local knowledge of natural resources management practices. With reference to these issues, SAGUN (a community development NGO) has undertaken this study to investigate the extent of knowledge pertaining to the environment as possessed by the Tamang residents a given area. The study entitled 'Tamang Indigenous Knowledge and its contributions to Forest Management and Participatory Development', carried out work in southern Kavre on the Mahabharat range of Nepal.

Government statistics claims that Nepal's forest area covers about 39 per cent of its landmass. In Nepal, all types of forests ranging from tropical and sub-tropical to temperate and alpine forests can be found. These forests support a rich diversity of flora and fauna, including a lot of endemic species. Unfortunately, it is recorded that Nepal is losing its forests at an average rate of 2 per cent per annum. Amongst many factors that contribute to such high rate of degradation is the disregard for indigenous forms of knowledge with respect to forest management systems. The following contribution is an enquiry into local aspects of Tamang indigenous knowledge of forest use and management based on ethnographic fieldwork.

The research sites

The case study was carried out in the villages of Budhakhani and Phoksingtar situated in Southern Kavre District on the Mahabharat range, which falls into the Central Development Region of Nepal. The Mahabharat Range stretches in a south-east to north-west direction and its elevation ranges from 500 m above sea level to 3500 m. Its width is estimated to be about 30 km on average (Majupuriaya and Kumar 1998). This range is covered with high altitude vegetation. Both Budhakhani and Phoksingtar are located on the southern slopes of the Mahabharat range covering 45.51 and 27.75 sq km respectively out of the total 1432.81 sq km area of Kavre District. The population composition of these two VDCs comprises as follows:

Population distribution by ethnic groups

VDCs	*Total population*	*Tamang*	*Magar*	*Bahun/ Chhetri*	*Occupational castes*
Budhakhani	3556	48 per cent	42 per cent	8 per cent	2 per cent
Phoksingtar	2199	75 per cent	20 per cent	1 per cent	4 per cent

Tamang cosmology

MYTH OF ORIGIN

The ecological significance of Tamang rituals must be examined in the light of their function as a set of rules that operationalises the production behaviour as a meaningful activity. Tamang rituals are fora through which the oral tradition is made socially significant. In Tamang society there are five types of ritual specialists: the *Lama*, the *Dorje Lobon* (clan deity priest), the *Tamba* (ethno-historian), the *Hyul Jyoho* (priest of territorial gods) and the *Benbo* (shaman). The Lama performs the death ritual, the Dorje Lobon performs the clan deity ritual, the Tamba recalls historical events and tells Tamang epics, the Hyul Jyoho performs the worship of clan territorial deities, and the Benbo performs shamanic rituals for the sick. Their authority for reciting myths through colloquial conversation or formal chanting is divinely sanctioned. In their collective role as knowledge-holders, these people are sources of cultural continuity who also bridge the gap between the spiritual and mundane aspects of the natural and social environment. Tamang cultural dispositions of nature and the relationship with the environment are manifested in their cosmovision, which is widely expressed through formal chants, various myths, legends, rituals, songs, stories, folk tales, and lore. *SAchhya-nAmchhya* and *Thungrap-kerap*, therefore, represent a mythical logic and explanation of the origin of the cosmos and its beings.

SACHHYA-NAMCHHYA — CREATION OF A WORLD

Kundu Sangbo (*Saminta Bhatta*)

Kundu Sangbo created the *sAchhya-nAmchhya* (earth and sky) through his eyes and consciousness in primordial times. The god created *namka* (sky) and *Dorje Gyadam* (cross) created the wind which filled the air above the fire and the universe. *Me Kilkhor Marpo,* the fire, created *Lungi Khorlo* (thunder and storm) and it started to rain. After the rain, the *Ling* (ocean) was formed and a typhoon (*Chhyuiki Khorlo*) made *Bobshya* (foam). The foam converted into *Dul* (dust) and the dust transformed into *Sa* (earth). From the earth, the *Rigyal Longpo Gang* (highest mountain) arose and subsequently, *Shergi Ridon* (seven golden mountains) also emerged around the *Rigyal Longpo Gang*. After that *Rolpe Chhodun* (seven oceans) were formed in the basins of these seven golden mountains. This popular Tamang account of the creation of a world is expressed in the song *sAchhya-nAmchhya*.

Luni Marpo and Luni Karpo

Two *Hoishers* (rays) occurred as *Luni Marpo Hoisher* (red rays of light) appeared from the east, and *Luni Karpo Hoisher* (white rays) appeared from the west. The *Luni Marpo Hoisher* created *Ngema* (the sun) and *Luni Karpo Hoisher* created *Lani*, the moon. After the creation of the earth and its environment, the *Narka Pang Dor Som* (three Dubos) came into being on the earth as the first plant and the *Pema Dharke Mhendo* (Simal tree) germinated.

Pema Dharke Mhendo (Simal tree)

From the *Simal* tree flower, five gods originated in all four directions at the centre of the Zodiac. *Lu Karpo* (white silver) originated in the east (*Syar*); *Sinze Sherpo* (yellow diamond) originated in the south (*Lo*); *Chan Marpo* (red copper) originated in the west (Nup); and *Jyang Mhamo Jyankhu* (green ruby) originated in the north (*Jyang*). Finally, *Guru Remporche* originated in the centre (*Wui*) as the heart of this realm. From the stem of the *Pema Dharke Mhendo*, the Simal tree, originated the bow, arrow and dog.

Thungrap-kerap (Origination/Birth of living beings)

The *Thungrap-kerap* is chanted or recited orally as both verse and prose during ritual ceremonies on special occasions by the *Lama*, *Dorje Lobon, Hyul Jyoho, Benbo*, and *Tamba* in the villages. In the beginning, there were no animals and plants and *Thungrap-kerap* (origination and birth) made living beings come into existence on earth, giving birth to new generations to continue life on earth. The *Thungrap-kerap* states that 84000 creatures were created in all four corners of the world.

Syar (the east)

Go Ngale Kewa (birth from eggs), 21000 reptiles were created by *Lu Karpo* in the water and spread all over the world from the east.

Lo (the south)

Lo Du Ngalne Kewa (birth from genital organs), 21000 mammals were created by *Sinze Sherpo* and spread all over the world from the south.

Nup (the west)

Doi Sherla Kewa (birth from heat), 21000 microbes were created by *Chan Marpo* from the fire and spread all over the world from the west.

Jyang (the north)

Juite Kewa (birth from air), 21000 plants were created by *Jyang Mhamo Jyangkhu* in the form of air and spread all over the world from the north.

Cheng-Resi (Ablotikeshor)

Ablotikeshor (Cheng-Resi) created *Phattewa* (*Jyang Chyum Shempa*) and *Maha Dasing Dolma* in the shape of monkeys. The couple gave birth to 500 animal children. These animal children tried to feed on their parents *Phattew* and *Maha Dasing Dolma* for their survival, so the parents went to *Ablotikeshor* who provided them with food seeds such as maize, wheat, barley and millet to be cultivated for their livelihood.

Hopame (Ami Tambawa)

Hopame (god) separated animals and gods/human beings by selecting six of the 500 animal children and made them into the six Buddhas. Who were called *Om Layathupa, Maha Laminthupa, Nie Miathupa, Pae Jyalsomthupa, Mae Hindathupa* and *Hung Ngalwathupa*. The *Om Layathupa* went to *Namka* (sky); the *Maha Laminthupa* went to hell, where living beings are slaughtered; the *Nie Miathupa* (*Mihlui Remporche*), being humans, settled on earth; and the *Pae Jyalsomthupa* went to the animal world in the forest. Similarly, the

Mae Hindathupa went to the evil world and the *Hung Ngalwathupa* settled in the under world. The mantra '*Om Ma Ni Pe Mae Hung*', therefore, is the incantation to the six Buddhas of the six worlds.

Song of Earth Re-creation

The Tamang conception of the re-creation of the world bears a significant meaning in terms of their relation to the forest and the sustainability of creation itself. One of the powerful oral sources of Tamang lore referring to this context states:

> At the end of time, seven suns and seven moons will shine. All the forests will dry up, stone and soil will melt away like wax and the earth becomes a lake. The god *Nangwa Thaya* will again create the earth, the green *Dubo* will grow and the *Tamda* will flower.

The *Tamda* is the primordial plant, which flowers first when the earth comes into being.

RITUALS

Doihi Chan (Clan deity) Ritual

The *Doihi Chan* (clan deity) ritual is performed for seven days. In this ceremony the community performs a series of rituals that re-tell the original myth of the Tamang. On the first day, the *Sildup* ritual is performed; the second day follows the preparation of *Nhor* (food items and other materials); and on the third day the hunting of the *Chiprong* (bird) takes place. On the fourth day the main ritual of the clan deity is performed, and on the fifth day the gathering of relatives is followed by a feast. On the sixth day the ritual is continued and the seventh is the day of farewell for the *Dorje Lobon*.

Sildup (purification of home)

The *Dorje Lobon* chants the *Serap* of *sAchhya-nAmchhya* (creation of earth and sky) and *Thungrap-Kerap* (origination and birth of living beings) in the *Sildup* ritual to purify the house.

Preparation of Nhor

Pokhri Geng (bread of rice flour), *Mase Geng* (bread of blackgrams), fish and other prescribed items are cooked. *Sangala Girma* (fermented millet) and *Mlala Bab* (fermented rice) are prepared and *Tangi la bam* (deer meat) is collected. *Shine* (unhusked rice) and *Mone* (rice) are made ready and *Syukpa Sang* (incense) is collected.

Hunting

In ancient times, the gods (*Lhot Gyalpo*) and humans (*Mih Mahnig*) used to talk with each other. *Dorje Lobon* was one of these gods and he is valued as equal to the gods when he chants the hunting myth *Mih Mahnig* and *Lhot Gyalpo* for hunting the birds *Chiprong* and *ShyaMihg* in the forest. Cashew-nut tree branches are placed on the ground and the birds are offered to the hunting god by sprinkling their blood. After completion of the hunting ritual, when *Mih Mahnig* reached his home, he threw the bird's legs inside the house and accidentally broke his wife *Gangsalmo*'s leg. Then *Lhot Gyalpo* suggested that *Mih Mahnig* should get married to another woman and gave him some seeds of maize, wheat, potato, and mustard to begin agriculture. Before departing, the *Lhot Gyalpo* and his daughter made *singara paasam* (fences) with cashew nut branches, *titepati*, peacock feathers, and *hyolwa* (a piece of cloth). Ensuring that the human being will no longer be able to see, talk and play with the gods, the hunter god left the earth behind and departed for the sky. Since then, the gods and man are separated. The gods can see and hear human activities but not vice versa. After the separation between gods and humans the *Mih Mahnig* remarried *Chhisalmo* and they took up farming.

The day of main ritual performance

To start the ritual, *La Tamba Remporche* will ask the *Dorje Lobon* to start the *Doihi*. *Dorje Lobon* starts the ritual of *Dohi Chan* with the help of *Jyoho Remporche* and a Lama who are associated in performing the ritual.

Doihi Pri

During and after the *Doihi Pri* ritual, the couple of the house and other members, should not cross the river or listen to bad words. Similarly, they do not quarrel, do not cut plants and trees, do not kill animals, do not dig the land, do not stay at night in somebody else's house, do not lend anything (in cash or kind) to others for 14 days. In the same manner, the female head of the ritual house should not listen to bad words, not quarrel, not cut or plant shoots, not kill any animals, and avoid staying at night in somebody else's house for a whole year.

Shergyam (ritual of territorial gods)

The *Shergyam puja* is performed to mark the full moon of Baisakh (April) by the *Hyul Dhami,* and the ritual involves the presence of all the tenants in the clan territory. This is another important ritual for the Tamang, and is performed to honour *Jyoho gang* (territorial deity) once a year when the Hyul Dhami is worshipped and appeased by offering fermented millet, wine, rice, and other food grains. During the appeasement, the Jyoho is requested to grant its benevolence to the clan and protect it from the unwanted forces of nature. After the *Shergyam*, a three day *Pri* (rest) is observed. During the three days of *Pri*, the villagers never listen to bad words, do not quarrel, do not cut plants and trees, do not kill animals, do not dig the land, and do not lend anything (either in cash or kind) to others.

Songs

Love song from the forest

The themes of the songs exchanged among the young girls and boys generally is about love, desire and aspirations. The young person's ability to share historical knowledge and use symbolic perceptions is also assessed during such occasions. One such song attempts to seduce the beloved one by saying

> Oh fairy-like beautiful lady, lips smiling like *Jyomo*, legs as smooth as the banana stem, where are you from? If you give me a little of your heart, I will come with my requests to your parents — let us have blossoms of flowers.

The song of 'ethnogeography'

The song of 'ethnogeography' has special ecological significance and makes references to the clan territory and its region. The song states that the Tamang territory extends from the base of *Glingla Che* (Himalayan peaks) to the Sailung hill range of Central Nepal. This supports the oral claim of the Tamang that the Kathmandu valley (*Yambu*) and its surrounding hills have been at the centre of Tamang cultural evolution.

Stories

Story of plants

There is a popular story in Tamang lore about plants and trees. Once upon a time, there were *Paata* (*Rhododendron*), *Bomsing* (*Alnus nepalensis*) and *Kyasing* (*Schima wallichii*) trees in the jungle. Once, *Kyasing* and *Bomsing* trees met at one place where *Kyasing* found *Bomsing* to be very handsome. *Kyasing* took *Bomsing* to the place of *Paata*. *Paata* was not flowering and looked very ugly, and therefore, was very much ashamed in front of *Bomsing*. The *Kyasing* and *Bomsing* returned to their own places. After some time, *Kyasing* came to *Bomsing* and took him again to the place of *Paata*. At that time, *Paata* was flowering and looked very beautiful. *Bomsing* proposed *Paata* to marry him, but *Paata* refused the proposal. After the failure of this marriage proposal to *Paata*, *Bomsing* went to the top of the hill and jumped down thus committing suicide. Since then, *Bomsing* is found only on steep slopes of hills and river streams.

Story of birds

Once upon a time, there were *Burgan* (pigeon) and *Mrawa* (peacock) in the forest. While Mrawa was very beautiful, Burgan was black and ugly. Burgan went to Mrawa and said,

> Tomorrow I have to go abroad, but since I am black and ugly, could you please lend me your beautiful red feet and take my black feet for the while. The peacock agreed to the proposal of the pigeon and they exchanged their feet. After that, the pigeon never returned its feet and always kept away from the peacock. Since then, the Mrawa always cries when looking at its ugly feet whereas the pigeon keeps singing when looking at its beautiful red feet.

Tamang perception of the environment is also significantly influenced by the origin myth, which is deeply intertwined with the forest. Different elements of nature have founded Tamang economy; established social identity; made them aware of political behaviour; and as a whole, formed the cultural model of their identity. In Tamang reasoning, the earth, sky, wind, fire, hills and mountains, water, rivers, sun, and moon are integral components of this identity.

Tamang myths of origin, stories, songs, rituals, and sayings indicate that the symbolic classification of the natural world is as follows:

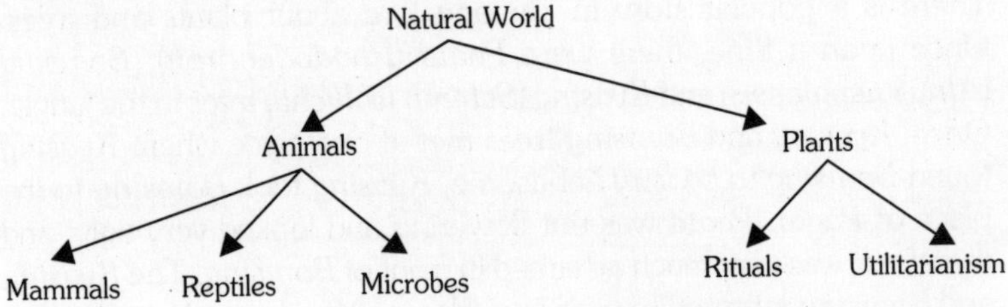

According to the origin myth, Tamang refer to *sAchhya-nAmchhya*, consisting of a number of mythological configurations. It includes the Tamang account of cosmological origin, its construction, the ordering of ethno-scapes and outer locations of the earth such as heaven, the under-world, human and extraterrestrial beings. The gods are able to help mankind in times of need to find the forests, and to make a living from them. The symbols such as the Chiprong bird, the deer, and the branches of a tree, are beyond comundane utility. Rather, these symbols recognise the mighty, co-operate realms of nature and the spiritual world. The significance of the forest in the creation and recreation of the cosmos appears to be extremely important to the conception of Tamang origin. Imaginations of the forest, its origin, and decay provides a very important clue to the relations of the Tamang with the natural world.

Whether it is through mythology, rituals, songs or folk stories, the Tamang have ample reasons to believe in the preservation and protection of the natural environment. One such example can be seen from the fact that after the practice of *Shergyam*, a three-day *Pri* (rest) has to be observed. During this period, villagers do not cut plants and trees, do not dig land and do not kill animals. This represents the need to preserve and protect the forest and other natural resources on earth for the continuity of its growth and regeneration in future.

Cultural meaning of resources

The above tales signify various underlying themes of the existence of a spiritual realm of gods that are not only supernatural but interact with human beings. Gods help mankind to find the forest for living, cultivate the land, and eliminating the sufferings of the body. The separation of nature and culture somehow gets blurred in the Tamang imagination whose 'super-nature' is effectively brought to the forefront where human beings, biophysical elements and supernatural forces unite.

THE SUN — SOURCE OF TAMANG TIME

The sun is a source of time and regarded as the master of all knowledge and awareness. The sun provides the light and guides humans to do things at the appropriate time. The position of the sun and degree of light provides the basis for understanding the time and season to use and manage resources. The natural processes around themselves are understood by the Tamang through the movement and state of the sun. According to the sun, they classify the seasons as *KhangbA-lA* (winter), *Serka-lA* (dry season), and *BarkhA-lA* (monsoon).

THE MOON — A SOURCE OF EXCHANGE

Lani (the moon) is considered as a resource as well as a source of exchange by the Tamang. A full moon night remind them to organise their cultural and aesthetic performances. Whether it is a ritual involving *Dorje Lobon* or a *Lama*, *Hyul Jyoho* or *Benbo*, the new moon or full moon days are auspicious days to worship deities. The full moon days of *Phalgun* (February), *Chaita* (March), *Baisakh* (April), *Bhadau* (September), and *Asoj* (October) are considered as special days to organise cultural events. Similarly, the new moon days of *Saun* (August), *Kartik* (November), *Mangsir* (December) and *Magh* (January) are very important for the Tamang to perform such celebrations. People gather at different places during these days for exchange and entertainment. Tamang boys and girls exchange songs and share feelings of life and hope.

THE EARTH — A RESOURCE TO LIVE

The earth is a resource and provides the basis of living for all living organisms, for agriculture, plants, trees, hills, mountains and oceans. The Tamang worship the earth in the *Shergyam* ritual, which is done with great respect. It is associated with a better yield of crops, good human and animal health, as well as the preservation of other resources such as forests and water.

THE WIND — BLOWS AWAY MISERIES

The *lhAbA* (wind) is generally considered as a blowing wind. According to Tamang belief, the *lhAbA* purifies the environment and blows away human miseries. A blessing, *yelA dukhA kAde lhAbA se BorgAi* (let the wind blow away all sorts of miseries) is one of the prominent blessings given by the elders. Without wind and air no living beings can exist in the religious belief of the Tamang.

PLANTS AND TREES — A RESOURCE FOR LIVELIHOOD

Tamang regard plants as one of the major resources (*Nhor*) for their livelihood and culture. They get food, clothes and shelter from plants. Besides the physical uses of the plants, there are many stories and tales associated with trees. For example, they regard big trees as the abode of the forest deity called *Dongbo* (tree deity).

ANIMALS — A RESOURCE FOR LABOUR, FOOD, AND AGRICULTURE

Domestic animals such as draught oxen are used for cultivation, as well as for meat which makes up a special part of Tamang food habits. Some animals are sacrificed to please different deities. They are also a very important source of cash income to supplement their livelihood. Animals play a symbolic role in exchange practices in the Tamang community: brothers often give animals such as goats, buffaloes and cows to their sisters.

The earth, hills, mountains, sky, wind, water, sun, moon, the plants and the animals are understood as *Nhor* (resources). These ideas reflect an indigenous philosophy that has emerged from Tamang culture and history. Each of the hills, trees, groves and places are associated with one or other legends combined with the utilitarian values of sentient beings. It is to be stressed that nature in Tamang religion is a great guide for the creation of meaningful social behaviour. The Tamang perceive the significance and cultural meaning of resources in both physical and psychological terms.

Tamang empirical knowledge

CLASSIFICATION AS AN ADAPTIVE MECHANISM

The resources of Tamang knowledge are produced and reproduced through the efforts of human beings, natural forces and spiritual agencies. Resources are classified into two forms. The skills and knowledge applied to change one form into another are called non-material resource, which the rest are understood as material resources. These types of resources and their combined applications can change during the production process. The principles of classification represent the cognitive domain which determine the resource selection.

The utility of natural components are understood through the classification of things being useful (*gyAt khelbA*) and non-useful (*gyAt aAkhelbA*). The Tamang perceives *sA* (the earth, land, territory, cultivating land), *sAprA* (soil), *dong* (tree), *rhi* (forest), *gAng* (hills), *Jyomo lung* (mountain), *lhAbA* (air), *lung* (wind), *syong* (river, rivulet), *kyui* (water) as core resources. Similarly, *mhe* (cattle), *mui* (buffalo), *rA* (goat), *nAgA* (chicken) *dwA* (pig) *rhi dwA* (wild boar) and other animals are also included in their list of resources. Having used this classification of resources over many generations, they have developed specific knowledge and practices over a certain type of land. However, the primary guiding factor for management is its utility depending upon diverse climatic conditions. Indeed, Tamang livelihood is based on diverse productive potentials of its physical environment. Local people commonly use these symbols to indicate different land features.

The various physical aspects of the Mahabharat Range makes the Tamang adopt diversified production strategies. The environmental zones in which people earn their livelihood covers a vertical elevation ranging from the tropical zone of 500 m at the Bagmati river confluence, to 3500 m at the peak of the Mahabharat Range. In the process of adaptation to these different zones, the Tamang community has developed a complex ecological knowledge to cope with diverse environments. In Budakhani and Phoksingtar VDCs, the practical experience of interacting with nature and the decisions made to use the environment are based on a cognitive mechanism. These perceptions are embedded in culturally-grounded attitudes and values. People earn their livelihood through sedentary agriculture and use the forest in various ecological resource zones. The landscape and topography of the hill ranges, soils and plants found at various altitudes, are perceived through a system of classification. Cultural rituals link beliefs to the empirical reality of their environments. The local people, through their system of nomenclature, distinguish landscapes and zones in the southern slope of the Mahabharat Range. Such systems help to describe the use and management of natural resources located at multiple elevations. In this connection, classification systems have become an adaptive mechanism for the Tamang community of the Kavre area.

Classification of Land Topography

The agroecological zones are divided into three distinct categories based on the principles of elevation. The Tamang classify topographic zones such as *lha*, *gung* and *lung*. *Lha* is an upper part of land where the climate is cold. *Lung* (low land) is the lower part of a hill where the climate is relatively warmer. The *gung* (central part of a hill) region is a zone of moderate temperature. Most of the Tamang settlements are found in this mid-temperate zone, but all three zones are used for their livelihood. Within these elevation zones, one can also find many different indigenous categories of the topography in the Tamang classification system such as *ble* (gentle slope), *brA* (steep slope), *gAng* (hill), *ko* (low pass), and *chhyaba* (plain).

LHA (UPPER ZONE)

Lha is the highest elevation zone consisting of areas above 1500 m and is in the temperate zone of the Mahabharat Range, which has two distinct features and falls under the category of customary land. The entire zone is covered with dense mixed deciduous forest and has perennial water sources. The forest is used mainly for fodder, animal bedding, medicine, building materials, fuelwood and pasture in the winter. In this zone, some Tamang communities exploit *bari* (rain-fed) land. A significant portion of the local economy rests on this zone. There are *chyulis* (peaks) in the Mahabharat Range near the Budhakhani and Phoksingtar VDCs. Close to these two VDCs there are three main peaks which symbolise the cultural significance of Mahabharat Range. These are :

***Jyoho chyuli*:** A story is associated with *Jyoho Chyuli*: Jyoho and the gods have an ability to shift their shape into sentient beings like animals, birds and human beings. One day, *Jyoho* said to his wife that he was going out hunting in disguise of a tiger. He gave her a few grains of rice and asked 'when I come back home, throw this rice over me then I will be again human'. But when *Jyoho* came home after hunting, his wife got scared to see a tiger in front of her and the rice (*mone*) slipped from her palm into the nearby fire. *Jyoho* became restless and tried to take rice from the fire. He burnt his right hand and transformed himself into a stone statue to overcome the problem. Since then, *Jyoho* wanders in the forest in the form of a white tiger. According to the villagers, there is a four-metre long tiger-shaped stone with a burnt right hand on top of the chyuli facing towards the north. All the villagers around *Jyoho Chyuli* worship the stone two times a year, once during full moon in *Baisakh* (April) and once in *Dashain* (October), through a sacrifice of goats and chickens. The villagers pray for protection from evil destruction and ask for prosperity in life.

***Bajradau chyuli*:** The name of this *chyuli* refers to the event of a big *vajra* (thunder), which struck a *Bepkar* (*Quercus* spp.) tree. The *chyuli* is known to indicate the unhappiness of deities in the area through such incidents. They cause heavy destruction in the forest. In order to prevent such natural calamities people appease the deities at this *chyuli* once every year.

***NupgAng gAng chyuli*:** At Budhakhani village, there are seven *gang chyuli*s such as *Nomang gang*, *Masle gang*, *Dursa gang*, *Chisar gang*, *Khundruk gang*, *Jamhe gang* and *Ghaiya gang*. *Nomang gang* related to the use of wild garlic in the past. In this area, Tamang do not eat garlic which not even allowed inside the house. It is believed that if some stranger brings wild garlic inside the village, the house deity will get angry and cause a heavy storm destroying the village. Moreover, it is also believed that it will bring clashes between the forest god and the god of the house.

GUNG: MIDDLE ZONE

Most of the village settlements are found in the mid-temperate zone, which is by-and-large the preferred elevation to exploit natural resources for the subsistence of the community. In this zone, only rain-fed agriculture is practiced. Almost all livestock rearing is done here using all the upper, middle and low-land pastures. Fodder trees for the livestock are grown around the farmland edges. Cattle, buffaloes and goats contribute to cash income with their meat and dairy products.

LUNG: LOWER ZONE

The sub-tropical zone of *Lung* (low-land) is called the basin of the foothills, which is relatively flat land. Half of the land is irrigated by using perennial water sources to grow paddy. There are very few houses here and cattle is seasonally kept to be fed and leave their manure to fertilise the fields.

The classification of the geographical setting comprising of different ecological zones is evident in the language used for referring to different parts of land within their territory. The indigenous categorisation of the agro-ecological zones in this environment is to some extent consistent with the scientific modes of classifying the topography.

Other dimensions of land classification

Local people assess and classify ecological zones based on their potential for agricultural growth. They consider the type of soil, wind exposure, and availability of water to assess possible crop yields. The classification of agro-ecosystems is of two types, namely *dipsa* (moist) facing the northern slope, and *thasa* (dry) facing the southern slope, of the Mahabharat range. They classify the hilly terrain in many forms according to its agricultural use and potential. The Tamang add a suffix or put a prefix to these words to indicate a particular place such as *ble-mrAng* (slope dry land), *ble-syong* (slope rivulet), *brA-Chyumber* (steeped slope). In this way, they have a particular term for each specific place.

CLASSIFICATION OF LAND

The land use types within the area are broadly classified into six major categories. The first is *Bu* (irrigated land) in the low-land (*Lung*) where Tamang cultivate paddy. The second is *Whara* (rain-fed land) most of which lies in the middle zone (*Gung*). The third is *Mrang* (marginal land), located at the edges of the forests. The fourth is *Sho Pakha* (roofing grassland) where the farmers grow different types of grasses for thatching the roofs of their houses. The fifth is *Charan* (pasture land) and *Lha Gothe* (upland grazing area) for cattle grazing. The sixth is Rhi (forest) and denotes both community and private forests.

***Bu (irrigated land)*:** Most of the *Bu* land is located along the river basin and other low altitude areas of the *Lung*. Paddy is grown during the monsoon, and wheat, maize, potato, and mustard in the winter.

***Whara* (*rain-fed land*):** *Whara* is rain-fed land found in the three zones of *Lha*, *Gung*, and *Lung*. Terracing slopes of the Mahabharat Range is a sophisticated undertaking of landscape modification for agricultural use. Crops such as maize, millet, wheat, potatoes, citrus fruits, and green leafy vegetables are grown here. The edges of the cultivated lands are used for fodder trees and fuel wood production.

***Mrang* (*marginal land*):** *Mrang* is usually found at the edges of forests, where ginger, *haldi* (turmeric), *ghaiya* (wet rice), legumes, *taya* (calocasia sp.), *teme* (yam), *sakar teme* (sweet potato) and chillis are cultivated. The *Mrang* supports additional food grains in subsistence agriculture.

***Charan* (*pasture land*) *and Kharka* (*high altitude pasture land*):** *Charan* (pasture land) is allocated for cattle grazing and plays a very important role in the economy as it provides manure for the household and fodder for the animals. *Kharka* (*Lha Gothe*) refers to temporary grazing land in the higher altitude of the Mahabharat Range, which is generally regarded as common property. The people of Budhakhani and Phoksingtar use the high-altitude forests of the Mahabharat as *Kharka* and keep their cattle and buffaloes there in the summer and during the monsoon.

***Sho Pakha* (*roofing grassland*):** Most of the farmers own at least a small plot for *Sho* (*Typha angustata*) grass for thatching.

***Rhi* (*forest*):** The land with tree cover, be it private or community-owned, is classified as *Rhi* (forest). It provides people with fodder and animal bedding, fuelwood, timber, and agricultural tools like plough, yoke and handles. Apart from these, the forests also support the community with various other things. For instance, incense (*bra mhendo, daling*), bamboo and vegetables such as *dekani, kundasali, chyali, pachyar polo, grang polo, bamboo shoots, pangkyung, kunardap, kastaya* and *kopi* are collected by the community in different seasons. *Shyamo* (mushrooms) such *as tanki, dishyamo, mharmo, naka shyamo, bepkar shyamo* and *maare shyamo* are collected in the forests. Fruits, like *karbisa, pulang, mhe pulang, hakupaku, chutra, kyakar palo* (*katus* seeds), and *lakar kyuanba katho* (oak) are collected. Tubers (*rhi teme*) and edible frogs (*peng*) are also used. Therefore, the forest is considered as a major source of livelihood for the community.

CLASSIFICATION OF PLANTS

Tamang have a vast knowledge of the life-cycle of a plant passing through different stages. The growth of plants from germination to ageing and death is described in a sequential order such as seed germination, stemming, shooting, sprouting of buds, flowering and seed bearing. They know that plants do not possess a more or less distinct shape like animals at the time of birth but that it is in the latter stages of growth, that new organs arise. Plants also make a significant contribution to their material and non-material culture.

The study identified over 519 plants including their multiple uses which are known to the local Tamang population. The folk classification system of plants has been mainly explored on the ground of their utilities, such as medicinal, ritual and ornamental. Furthermore, plants are used for fodder, fuel, food, timber, soap, cloth, paper, dye, rope, and agricultural tools. The plant species are categorised based on morphological characteristics. The villagers distinguish different classes such as *dong* (tree), *jhang* (bush), *pothra* (shrubs), *gugi* (climber), and *mra* (weed) according to the rooting structure, height, density, shade potential, location and elevation of the location. Apart from the use of plants for economic and productive purposes, the social context and cultural way-of-life makes use of trees and plants.

CLASSIFICATION OF WILDLIFE

The classification of wildlife is based on Tamang hunting practices. Knowledge of hunting is based on the experience of ancestors and subsequently inherited by the younger generation through rituals and oral traditions. Although hunting is no longer of practiced, the evocation of hunter gods and wildlife-related mythology and rituals have kept the Tamang imagination and culture alive. Animals are broadly classified into two categories such as *Shya chapa* (carnivorous) and *Shya Achapa* (non-carnivorous). For instance, the carnivores are *Chyan* (tiger), *Tawar* (cat) and the wolf. Herbivores are the *Tanki* (deer) *Dumsi* (porcupine), *Rhi dwa* (wild boar), and so on. They also classify wildlife as *sA Kewa* (terrestrial living beings) as opposed to *Kyui Kewa* (aquatic living beings).

Power politics and forest management

Indigenous systems, as argued by Gilmour and Fisher (1991), are not only a remnant of tradition, nor are they simply locally-accepted modes of practices which are favoured by formal authorities. Rather, indigenous management systems comprise a dynamic response of the local population to their changing needs influenced by outside forces, internal dynamism and environmental contingencies. In this way, Central Nepal has been the focal point of historical power struggles to establish control over the fertile land of the valley and its surrounding hills. Armed forces from outside invaded the Tamang settlement area and its natural resources began to recede with the political unification of Nepal under Prithvi Narayan Shah. On this background, the history of forest management can be broadly categorised into four stages.

THE FIRST STAGE (1768 TO 1846)

The first stage begins in the late eighteenth century and includes the period of early Shah rule in the Kathmandu valley from 1768 to 1846. After the unification of Nepal, the Tamang socio-political set-up was brought under Gorkhali rule. The time of unification from 1768 till the beginning of the Rana regime in 1846 is a phase of selective exploitation of Tamang settlement areas. At this time, most of the forests in Tamang territory were protected exclusively for making gunpowder, charcoal, fuel-wood and timber for the government. During the war of 1814-16 with the British East India Company, it was ordered to remove settlements south of the Mahabharat Range and north of the Siwalik range area, to close all tracks, and to let this area revert to forest. Local people were asked to plant bamboo, cane, and thorny bushes to render the tracks unusable (Gautam 1991:31). Check-posts along the central part of Mahabharat Range were introduced in 1817, 1824, and 1826. The history of the Mahabharat Range forests have gone through various changes leading to the abandonment of indigenous forest management systems.

THE SECOND STAGE (1846-1950)

The second stage started from the intrusion of the Rana regime, and lasted for over a hundred years (1846-1950). This period can be taken as a dark, turbulent political period for forest management practices. The Rana regime, having absolute power over the country, engaged in the exploitation of resources for their own benefits through drastic depletion of forest resources (Bista 1991). A new land tenure system was introduced that completely destroyed the *kipat* system of communal land ownership. In the hills, a new taxation system operated through mechanisms such as the *Talukdar* system, where a landlord was usually appointed for revenue collection. The traditional forest management rights of the local people were transferred to the tax-collecting agent of the state. The interest of the government was in this way to clear the forests for agricultural land, and to raise more rent to finance its huge state expenditures. In addition, with an objective to supply the Rana nobility's dairy products, swathes of forest land were converted into pasture. Cattle and buffaloes were raised at a large scale, which significantly contributed to forest depletion.

THE THIRD STAGE (1951-80)

This stage began after the dismissal of the Rana Regime in 1951, until formal rules and regulations for forest management were enacted by the local communities under the leadership of the Pradhan Panch (locally-elected village government) in 1981. For ten years, the democratic Panchayat system and the king co-existed until 1961, when King Mahendra Bikram Shah proclaimed full power of the Panchayat system. The Government declared the Forest Nationalization Act in 1957 as a new centralised measure to protect the forest. By this Act, all forests were declared government forests and local users lost their ownership. The Panchayat System enacted a Land Reform Act in 1964 and the status of *Talukdar* was abolished. Most of the public arable land of the area was then registered as private land by the local elites in the cadastral survey of 1977. In the period from 1964 to 1978, it has been estimated that the amount of forest area with more than 10 per cent crown cover per hectare has fallen from 40 per cent to 13 per cent in the middle hills and mountain areas (Adhikari 1990). However, the Act failed to protect the forests and another National Forestry Plan was formulated in 1976 that allowed possibilities for converting nationalised forest into Panchayat forest. People living inside the political boundary of a Panchayat collectively managed and used the forest resources until the concept of community forestry was introduced following the establishment of party democracy in 1990. However, the situation of the forests worsened during this period and the livelihood of many communities deteriorated further. The disappearing forest vegetation and increasingly barren slopes caused heavy land-slides and soil erosion in the 1970s.

THE FOURTH STAGE (1981-98)

The first three stages outlined above were periods of forest destruction and a complete failure of the national forest policies adopted by the state. Eventually, after thirteen decades of external control, the management and control of forests returned to the local level after 1981. Community forestry was recognised as the most significant strategy for the fulfilment of people's needs by the Master Plan for the Forestry Sector in 1988. During this period, rights of communities over forest use were granted, and the Community Forest Management Programme was introduced. Keeping in view, the issues of ownership and rights to the land, a power struggle and resistance has evolved in Tamang cultural politics. The distribution of forest resources are some of the key themes that govern and guide forest management in Nepal as an emerging modern nation state.

Community forestry

In the two villages of Budhakhani and Phoksingtar, forest management committees were formed under the leadership of a Pradhan Panch (elected chairman of the Village Panchayat) in 1981 for two reasons. Firstly, the successive catastrophic floods and landslides of 1954 and 1971 greatly generated awareness among the local people to prevent such disasters in the future. Secondly, the simultaneous loss of forest through human intervention and natural calamities raised serious concern for the problems of fodder supply and firewood collection. Since this time the participatory Community Forestry Management Programme has developed through the formation of user groups and the introduction of management plans.

In Budhakhani VDC, *Mahuwa-ghari, Chautara Pakha, Kosing whang Pakha, Dhungkhure Dursa gang, Jyoho Chyuli, Bage Bhanjayangt,* and *Wati Pak. a* are the community forests. In Thulopokhara and Phoksingtar VDC, *Mhe Gothe Pakha, Dursa gang, and Gairiban Pakha* consist of community forests. All these forests are located on the southern slopes of the Mahabharat ranging from 800-2200 m, and are managed by communities through the formation of forest user groups. According to the community members, all these forest areas were once covered by dense natural vegetation with a rich bio-diversity of plant species. Political influence had a remarkable impact on the degradation of the forest resources, and most of the present patches are secondary forest, except for the top-most forest in the Mahabharat Range.

From the situation stated above we can summarise that the utilisation of forests and their management in this area is influenced by four major factors: the indigenous social organisations that control the access to forest resources; the customary norms and procedures to control and allocate resources; the concept of community forestry in accord with the environmental perception of the Tamang people; and, lastly, the significant contribution that can be made to the subsistence of the local community. In traditional forest management systems, people can clearly make rational management decisions through socio-cultural mechanisms such as sacred sites, which are effectively conserved. However, when traditional rights are infringed by an external authority, resulting in the loss of ownership this, often manifests itself in destructive attitudes that cause negligence in conservation.

The forest, upon which the livelihood system of local people largely depends, cannot be conserved without their active participation, as is evident from the enormous loss of forest during the centralisation period. In the past, when the government experimented with various models to solve the forest crisis, traditional or indigenous management systems were ignored. When local decision-making power was removed, management problems surfaced, and could not be effectively solved until local ownership was acknowledged due to active participation of the local community. Other major features of indigenous management are:

- The feeling of landownership and inalienable rights to the land as a key feature of its proper management. This feeling is continuously reinforced through cultural beliefs and rituals.
- Depending upon the situation, people let the forest grow for different periods of time, sometimes for one season and sometimes for longer. Meanwhile, important sites such as water sources, resting places, spiritual and religious places, are conserved.
- The villagers adhere to the principle of using forests according to their needs such as fuel-wood and fodder collection, building materials, poles, and so on.

These needs are determined by socio-economic factors but forest resources can be distributed unequally. Sometimes, poorer families have greater dependency on common property resources but may not be able to extract and consume as many forest products as the richer households in the same village.

Conclusion

The overall perception of the environment is based on both environmental knowledge as well as the human relationship with the supernatural. In Tamang society, a network of rituals and symbolism transforms the mundane into a set of meaningful actions. Traditional management systems and culture are therefore pivotal. Local efforts with common ownership rights will lead to a satisfactory policy, whereas there is little prospect of success if forestry relies on enforced regulation and ultimate government ownership. There is a danger that the formalisation of community forestry and forest user group certifications issued by the government which have no cultural value will only be perceived as technical.

Local management systems, as can be seen from the above cases, can survive if there is little external intervention. Numerous benefits may arise if the government decides to facilitate community systems wherever possible. This could symbolise an end to the policing role by the government and help to increase or maintain the power of village leaders as mediators between local people and the government authorities. The new collective exercise of forest conservation can produce results, such as gradually improving barren land into patches of forest. Yet, forests can only be managed through the decision-making authority of the community once their land rights have been given back to them. Active community participation is possible if based on a fair contribution of labour input and a distribution of benefits among the members of a forest user group. In general, the degradation of natural resources is a result of complex interrelated factors and historical processes. Development strategies that incorporate cultural explanations of the human-environment relationship must be examined on this background, rather than simply forming forest user groups in a technical sense for the empowerment of local people. Since local people depend to a great extent on forests for their livelihood, the conservation of forests cannot and should not depend upon isolation. It is therefore extremely necessary to further strengthen participatory community forest management for a truly collective vision of effective forest conservation.

References

Adhikari, J., 1990, *Is Community Forestry a New Concept? An Analysis of the Past and Present Policies affecting Forest Management in Nepal*, Society and Natural Resources, Kathmandu.

Bhattachan, K.B., 1993, 'Nepalka Adibasiharu: Abhidharna Ra Pahicha'. A paper presented at Rastriya Gosthi, Social Council, Sociology/Anthropology Dept. of Tribhuvan University, Kathmandu.

Gautam, 1991, Indigenous Forest Management Systems in the Hills of Nepal, M.Sc. Thesis, Australian National University, Canberra.

Gilmour, D.A., Fisher, R.J. 1991, *Villagers, Forest and Foresters : The Philosophy, Process and Practice of Community Forestry in Nepal,* Kathmandu: Sahayogi Press.

Majupuriaya, T.C. and Kumar, R. 1998, *Wild Life, National Parks and Reserves of Nepal*, Bangkok, Craftsman Press Ltd.

Regmi, M.C., 1978, *Land Tenure and Taxation in Nepal*, Kathmandu: Ratna Pustak Bhandar.

Tamang, P. B., 1994, *Syo Mhendo,* vol. 17, Kathmandu. Tamang Gyai Den Sengleng Senjyang.

15

Traditional Use of Blue Pine (*Pinus wallichiana*) and its Impact on the Forest Structure in Central Bhutan

Christina Giesch

Introduction

THIS paper is part of a PhD research project on the evolution of traditional forest uses and their impact on the forest structure, with reference to issues of sustainability. Knowledge of traditional forest activities, their rationale and impact on forest structure is necessary for a sustainable forest management taking the needs of the local population and their indigenous knowledge into account. Blue pine (*Pinus wallichiana*) is by far the most commonly used tree species in Bumthang District in Central Bhutan. The blue pine forests are located at a relatively low altitude, close to the valley bottoms and settlements, and are therefore easy to access. Furthermore, the wood from blue pine is used for many purposes and is essential for the construction of houses. Traditionally, the harvest of blue pine trees follows a rationale which differs considerably from that of professional foresters. This paper describes the traditional use of blue pine and the impact of blue pine extraction on the forest structure. In Bumthang, forest covers 50 per cent of the district area and is steadily increasing, whereas forests is decreasing dramatically in other parts of the country. In southern Bhutan, 13,600 ha of broadleaf forest disappear every year (Wangchuk 1998). Consequently, the results of this study and its conclusions are limited to Dhur in Bumthang, the study area to which this account refers.

Methods

The research combined a sociological approach in different village communities and a forest inventory. A household survey was made with semi-structured interviews during the winter of 1996-97 covering all households of the research area. This survey documented the socio-demographic situation, the economic conditions, activities and historical aspects of the area.

During the winter of 1997-98 a two-phase forest inventory was carried out in order to map the forest structure in the study area. In the first phase qualitative data (mainly the description of the forest and forest use) was collected, and in the second phase the dendrometric data for trees and stumps were recorded. A total of 800 ha of forest was taken into account with a rectangular grid of 200m × 200m. The inventory gives valuable information about the type of trees used and the impact of harvesting on the forest structure. A forest resource use survey also provided detailed information on the availability, consumption, mode of procurement, sale and future outlook of the local population on the management of timber and non-timber forest resources.

Location, natural and social setting of the study area

Bhutan can be divided into three geographical zones: the Southern zone (altitude: 200-1,000 m); the Central zone (1,000-4,000 m); and the Northern zone (above 4,000 m). The main valleys are oriented north to south and separated by ridges reaching up to 5,000 m. Bumthang is located between 27°30' N and 27°45' N latitude and 90°35' E and 91°55' E longitude. It comprises parts of the Central and Northern zones: the lowest altitude in this area is approximately 2,500 m and the highest 5,700-5,800 m (Land Use Planning Project, LUPP 1994).

Bumthang is one of the twenty *dzongkhags* (districts) in the country and has a total area of 2,714 sq km (LUPP 1995). It consists of a complex of four comparatively broad valleys (see *fig.* 1). Each valley is a *geog* (sub-district): Chhoekor, Chhumme, Tang and Ura. The study area, Dhur, is located in *Chhoekor Toe*, the northern part of Chhoekor *geog*. Dhur is the second biggest village in Bumthang

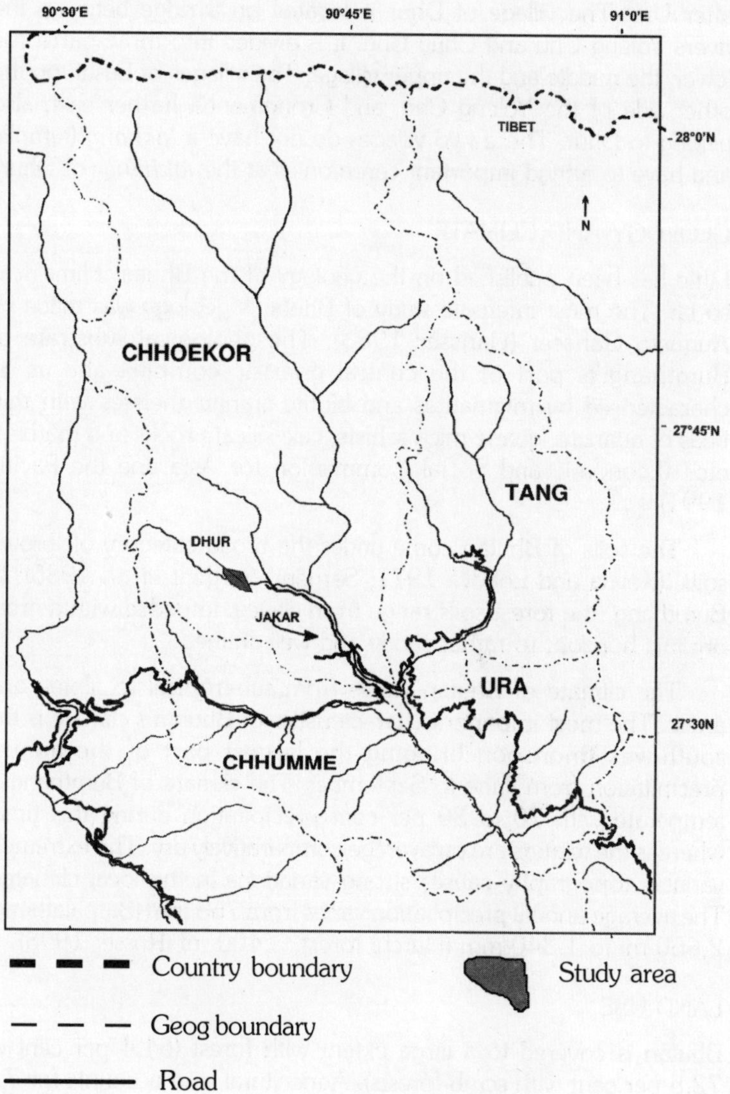

***Fig.*1 : Map of the Bumthang district. 1:100,000. Modified from LUPP 1995.**

after Ura. The village of Dhur is located on a ridge between the rivers Yoleng Chu and Chuti Chu. It is divided into three parts: the lower, the middle and the upper village. The villages of Lusbi, on the other side of the Yoleng Chu, and Crongmanba further east, also belong to Dhur. These two villages do not have a *lakhang* (temple) and have to attend important ceremonies at the *lakhang* of Dhur.

GEOLOGY AND CLIMATE

Little has been published on the geology of the Bhutan Himalayas so far. The most intensive study of Bhutan's geology was made by Augusto Gansser (Gansser 1983). The geological substrate of Bumthang is part of the central gneissic complex and as 'is characterised by migmatites and biotite-granite-gneisses with thin beds of quarzite, quartz-mica schists, calc-silicate rocks and marbles, etc.' (Economic and Social Commission for Asia and the Pacific 1991).

The soils of Bhutan come under the broad category of 'brown soils' (Kovda and Lobova 1971; Sargent, Sargent *et al.* 1985). In Bumthang, the forest soils range from clayey, impeded with a great organic horizon, to rather sandy and well-drained.

The climate of Bhutan varies from sub-tropical to alpine and arctic. The most important characteristics of Bhutan's climate is the south-west monsoon bringing the largest part of the annual precipitation from June to September. The climate of Bumthang is temperate with 70 to 80 per cent precipitation during this time, whereas the winters are on average comparatively dry. The extremely variable topography causes strong variations in the local climates. The average annual precipitation varies from 763 mm (Batpalathang, 2,650 m) to 1,340 mm (Hurchi forest, 3,400 m) (Rosset 1998).

LAND USE

Bhutan is covered to a large extent with forest (64.4 per cent or 72.5 per cent with scrub forests). Agricultural land accounts for 7.7 per cent and pastures for 3.9 per cent of the country's territory. The distribution of land types in Bumthang is presented in Table 1. The forest covers half of the district, which natural pastures (8 per cent) and unproductive land (23 per cent) are above the national average, owing to the fact that a large part of Bumthang is above the timberline.

Table 1: Land type

Land type	*per cent of the total area*
Agricultural land	2
Forest	50
- Fir	(15)
- Blue pine	(8)
- mixed	(27)
Scrub forest	17
Pasture	8
Others (settlements, rocks,...)	23
Total	100

Source: LUPP 1995.

Agricultural land makes up only 2 per cent of the district's area. *Pangzhing,* a from of grass fallow under shifting cultivation (Roder, Calvert *et al.* 1992; Roder, Calvert *et al.* 1993), is the most widespread type of agricultural land (61.5 per cent) used for the cultivation of bitter and sweet buckwheat (*Fagopyrum tataricum, F. esculentum*) which was the staplefood until some time ago.

FOREST TYPES

The forests of Bumthang are temperate coniferous forests, some with a rich broadleaf understorey. The main species are (from lower to higher altitude): *Quercus semecarpifolia*, *Pinus wallichiana*, *Picea spinulosa*, *Tsuga dumosa*, various *Rhododendron sp.*, *Betula utilis*, *Acer sp.*, *Juniperus sp.*, and *Abies densa*. Larch occurs in Ura, but was not found in the study area.

Table 2: Distribution of forest types in Bumthang

Forest type	*area in ha*	*per cent*
Broadleaf (>80 per cent)	33	< 0.1
Broadl./conifers mixed	15	< 0.1
Blue pine	22,015	16
Mixed conifers	72,175	54
Fir	40,438	30
Plantation	9	< 0.1
Total	134,685	100

Source: LUPP 1995.

Table 2 shows the different forest types, of which the most prominent is the mixed conifer forest with 54 per cent and comprises of fir, hemlock and spruce. These forests are at comparatively low altitudes near the valley bottoms. Broadleaved and mixed broadleaved forests, conifer forests, and plantations are almost lacking.

BLUE PINE FORESTS

The Himalayan Blue Pine, *Pinus wallichiana* A.B. Jackson (*P. excelsa Wallich ex D.Don*, *P. Griffithii McClell*), is a pine with its needles in clusters of five, and is found from Afghanistan to south-east Tibet [Polunin and Stainton 1997]. In Bhutan, it is commonly found at altitudes ranging from 1,700 m to 3,300 m (Grierson and Long 1983). As climax forest, it is found on dry sites (south exposition and ridges) in pure stands and mixed with other species in moister places. It is also a pioneer species which colonises fallow lands vigorously (Brandis, 1906; Champion, 1968; Grierson, 1983; Troup, 1986 rep.). For example, Sangay Wangchuk (1998) indicates that in the neighbouring *geog* Chhumme at least 33 per cent of the agricultural land was invaded by blue pine.

In the inventoried forest for this study, blue pine forests and blue pine forests with an oak understorey represent 28 per cent of the area or 220 ha. They are found in the lower part of the inventoried forest at 2,700 m to 3,300 m. Pure blue pine forests were found mainly at the fringe of pastures and on the southern slopes of small ridges. On average, blue pine forests had 752 stems per ha (9 per cent relative standard error), a mean basal area of 29 sq.m/ha (rse: 10 per cent), and a mean volume of 339 cubm/ha (rse: 12 per cent). These data are similar to those of Ernst Rohrbach (Rohrbach 1989). Blue pine forests with an oak understorey were found on more mesic sites. They differ from each other mainly because of a higher proportion of broadleaves and number of stems per ha. They had on average 937 stems per ha (rse: 8 per cent), a basal area of 39 sq.m/ha (rse: 10 per cent), and a volume of 471 cubm/ha (rse: 14 per cent).

Social and economic conditions

POPULATION

In 1993 Bhutan's population was estimated to be approximately

640,000 (Central Statistical Organization 1994). This is a population density of 15.7 inhabitants per sq km which is very low as compared to those of the neighbouring countries. No official figures are available for Bumthang's population, but it can be roughly estimated to be 1,300 households and 12,000 inhabitants. Its population density (4.4 persons/sq km) is less than the national average. Since the large northern part of the district lies above an altitude of 3,000 m, settlements can be found only in the southern part of the Dzongkhag. The only urban centre of the district is Jakar in Chhoekor geog.

In the study area 82 households with 622 people were recorded during the winter of 1996-97. The male-female ratio is almost equal. The distribution of the population of the study area according to social groups and gender is shown in Table 3.

Table 3: Social groups and gender in the study area

	Monpa	*Brokpa*	*Others*	*Total*
Male	161	134	18	313
Female	167	129	13	309
Total	328	263	31	622

Source: Household survey by the author, 1996-97.

For centuries, two social groups have been settled in Dhur: the Monpas (or Mon) and the Brokpas. After the Chinese invasion of Tibet, many Tibetan refugees migrated to Dhur and shifted to Jakar later on to start a business. Only few people from other regions settled down in Dhur. They either came for a particular reason or to do a specific job (schoolteacher, forest contractor...) or married a 'Dhurpa' (native of Dhur). Every summer about 4 or 5 families come arrive from Kheng (Zhemgang District) in June to graze their cattle in Dhur and go back in September-October.

The Monpas consider themselves the indigenous settlers in this region (*Chhoekor Toe*). The term Mon, or Monpa, is confusing, however, as several groups called Monpa can be found in various places: in Bumthang (farmers), in Kheng (forest dwellers) or in Arunachal Pradesh, India, where they are distinct ethnic groups. The confusion is a result of the original meaning of this term: 'Mon is a general term used by Tibetans for many Himalayan regions inhabited by non-Indo-Aryans' (Aris 1994). The Brokpas believe they originate from Tsampa, the border region between Bhutan and Tibet, and fled from this area due to the numerous wars during the reign of

Shabdrung Nawang Namgyel in the seventeenth century. The Brokpas of Choekhor Toe still speak their original language (*brokkat*) and have their own customs. George van Driem (Driem 1993) listed brokkat as one of the Bhutanese languages with approximately 300 speakers.

ECONOMY

The majority of Bhutan's population lives in rural areas as subsistence farmers. The GDP (Gross Domestic Product) per capita of 478 US$ in 1995 (UNDP 1996) is relatively high as compared to the neighbouring countries. The main contributors to the GDP are agriculture and forestry (43 per cent in 1993), electricity (8 per cent), transport (8 per cent), and manufacturing of goods (9 per cent) (World Bank 1994). The energy sector is expected to grow rapidly with the construction of several hydropower plants. The situation in Bumthang is similar to the Bhutanese average: the vast majority of the population are subsistence farmers or herders. The livelihood of most Monpas is based on subsistence agriculture and the Brokpas are mostly semi-nomadic and make a living from their Yak (*Bos grunniens*) herds. The occupation of the inhabitants of the study area and their residential status is shown in Table 4.

Table 4: Main occupation and residential status of the active population in Dhur

Occupation	*Permanent*	*Semi-permanent*	*Absentee*	*Total*
House-wife	125	3	15	143 (27 per cent)
Student	66	3	44	113 (22 per cent)
Herdsman	10	53	37	100 (19 per cent)
Farmer	71	1	1	73 (14 per cent)
Monk/Priest	21	1	16	38 (8 per cent)
Others	15	2	37	54 (10 per cent)
Total	308 (60 per cent)	63 (12 per cent)	150 (28 per cent)	521 (100 per cent)

Source: Household survey by the author, 1996-97.

22 per cent of the active population is enrolled in schools, a large part being in or near the village. Pupils of the higher classes must leave their village to join boarding schools. A considerable number of men are monks living in monastic communities and 10 per cent are engaged in tertiary economic activities, mostly outside the village.

INSTITUTIONS

The *dzongkhag* (district) is administered by a *dzongda* (district administrator) and his assistant the *dzongrab*. The *dzongkhag* is divided into several *geogs*, which are headed by a *gup*. The *gup* is elected by the villagers and is the link between the villages and the *dzong* (administration). The *chhimi* is a national assembly member representing a *geog* and elected by the villagers of the *geog*. The representative of a village is the *tshogpa*. He is elected by the villagers for a term of 3 years, and is exempted from work for the government and compulsory village work. The *Chipon* or liaison man is a messenger who keeps the contact between dzong and the villages. In the study area, each household delegates a liaison man for a one year term. Furthermore, every temple has a caretaker, *gonyör*, who is elected by the villagers.

INFRASTRUCTURE

Until the late 1950s, Bumthang was totally isolated and could be reached only on foot or with a horse or mule. Infrastructural development started in the 1960s with the construction of the national west-east highway, schools and the establishment of health services. The national highway crosses the district from west to east with 122 km of metalled road. A hydroelectric power-station in Chhumme with a capacity of 3 × 0.5 Megawatt provides electricity to parts of Chhumme and Chhoekor *geogs* (Central Statistical Organization 1994). At the end of 1992, telephone and fax communication was introduced complementing the post office and wireless facilities.

A forest road leads to a suspension bridge near Dhur, from which one has to walk 15-30 minutes to the villages. From Dhur start two mule tracks: one leading to Trongsa (about a four-day track) and one to Dhur Tsachu (hot springs, about 2-3 days) although this track has been officially closed since the Chinese invasion of Tibet. From Lusbi there is a footpath to Lugirawa gompa (monastery), Khole gompa, and finally crosses a pass to the neighbouring valley. From there a 3-4 day mule track leads to Tibet. The village has a community school, a basic health unit and neither electricity nor a telephone connection.

THE FOREST ACT 1969

The Bhutan Forest Act was promulgated in 1969. With this Act, all the forests to which 'no person has acquired a permanent, heritable and transferable right of use and occupancy' (RGOB 1969), were declared government reserved forests. The traditional use was restricted (e.g. the felling of trees required a permit and marking) or even forbidden (e.g. the burning of pastures or the collection of torchwood). The Forest Act guarantees the access to timber for bonafide use for the local population against payment of a royalty.

Traditional use of blue pine

In this context, 'traditional' is defined as 'existing before modernisation took place in the late fifties'. Modernisation and new laws such as the Forest Act of 1969 brought about rapid changes, particularly for the rural population. As described above, blue pine is commonly found in Bumthang, where it has a share of over 16 per cent of all forests. The wood of blue pine is soft and easy to work with. When dry, it has a relatively low weight which is an important aspect, since the terrain is difficult and pack animals are not used. All timber such as beams and planks as well as firewood has to be carried. Heavier wood species would make the use of wooden products in these quantities almost impossible.

Fuelwood

The households in Dhur depend fully on wood as a source of energy. Only one household uses a gas stove for cooking, but still needs wood for heating and to bridge the gap when the gas canister is refilled (which might take up to several weeks). The villagers therefore ranked fuelwood as the most important forest product in the household survey. While fuelwood needed daily, construction wood is needed only from time to time for repair works or new constructions. In the past, the villagers distinguished two types of fuelwood: *jarshing* (summer fuelwood) and *günshing* (winter fuelwood). For *jarshing* men felled one or more trees and chopped them. The wood was left to dry in the forest and women were to carry the dry wood to their homes. For this work, the women of one household requested women from other households to help them carrying their loads. As remuneration, they offered meals and wine to their helpers. The same mutual aid can be observed with other works: house construction, replacing the shingles on the roof, and digging the buckwheat fields. *Günshing* is mainly dry wood collection from small trees. Nowadays, the forest service issues permits for fuelwood and marks the trees once in a year and the wood is free of cost. For the Brokpas, this system has disadvantages since they are not always in the village and might miss the day on which the forest service man comes to the village, and thus may not get trees marked for their fuelwood supply. The mean diameter of blue pine trees that

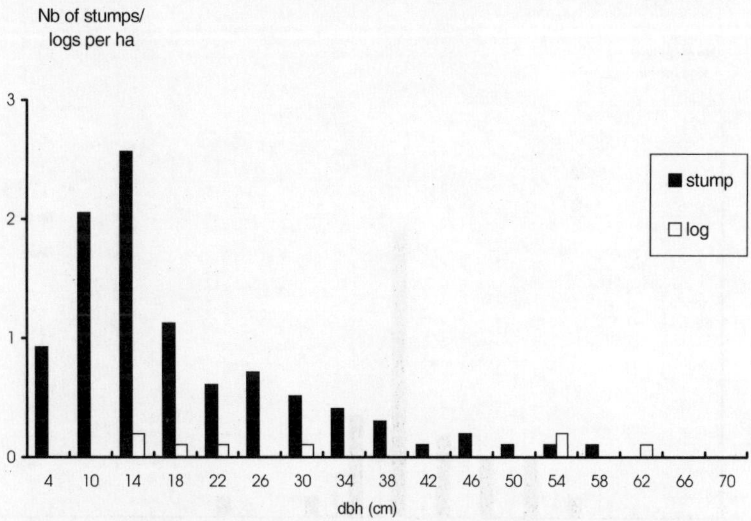

***Fig.* 2: Diameters of blue pine stumps and logs felled for fuelwood.**

are cut for fuelwood is 12 cm and the range 2 cm to 60 cm. The distribution of the diameters on blue pine stumps and logs of trees cut for fuelwood is shown in *fig.* 2.

Construction wood

Households planning or doing some construction work ranked the availability of construction wood as important, whereas households which were not concerned with any repair or construction work ranked it medium important to unimportant.

BEAMS AND PLANKS

Chams is the local name for beams. They are essential for house construction and preferably made from blue pine. In rare cases, when blue pine is not available, other coniferous species such as spruce or fir are used. A *cham* is produced by chopping off the round edges of a stem and thus out of one stem one *cham* is made. Depending on the height of the tree, two to three *chams* can be got from a single tree. The tree chosen for chams has to be straight and preferably without branches. The mean diameter is 30 cm (ranging from 20 to 44 cm, cf. *fig*. 3). According to statements from the villagers, a tree

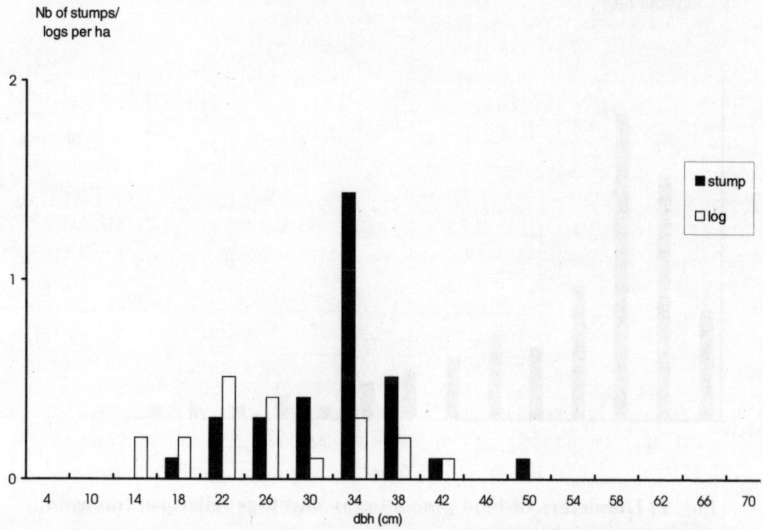

***Fig.* 3: Diameters of blue pine stumps and logs felled for beams.**

for *chams* has to be felled during the first 19 days of a month (lunar calendar). The chopped *chams* are left in the forest to dry. Once dry, they are carried on the back to the village.

In the past, only few Brokpas owned houses in the village. Nowadays, almost every Brokpa family has at least a small house. More and more Brokpas are giving up Yak herding and rearing to settle down in the village so that their need for construction wood is now similar to that of the Monpas.

PLANKS

Planks are made of blue pine, fir and spruce. Traditionally, planks were usually chopped with the axe. Nowadays, some carpenters use saw pits to saw the planks. Two planks could be made from a part of a stem with an axe; with a saw one can get three to four. The advantage is clearly seen by the villagers and sawn planks are readily accepted. The villagers living close to a road, however, do not have a choice: they have to buy the planks from a sawmill and are not

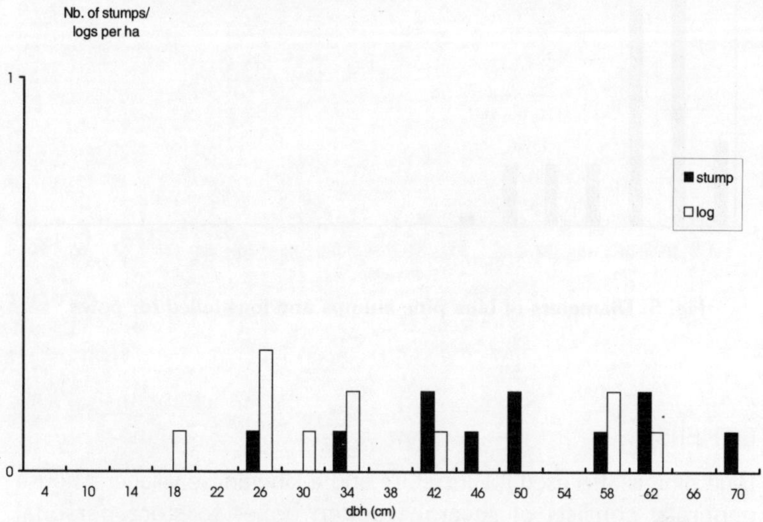

***Fig.* 4: Diameters of blue pine stumps and logs felled for planks.**

allowed to chop them themselves. Trees with a large range of diameters are used for planks, but in general these trees are bigger than those used for *chams*. The mean measured diameter of trees used for planks is 51 cm, ranging from 27 cm to 84 cm (*fig.* 4).

POLES

Poles are used for various purposes: for fencing, tying the cattle and Yaks, or for prayer flags. For domestic purposes, different species might be used, whereas poles for prayer flags require straight, slim and long poles, hence young conifers. The mean diameter of trees used as poles was 12 cm (see *fig.* 5). The Monpas mainly need poles to fence their fields in order to prevent cattle and game to damaging their fields and crops. The Brokpas need poles primarily to tie their animals to, and to construct temporary huts while they are camping on their pastures.

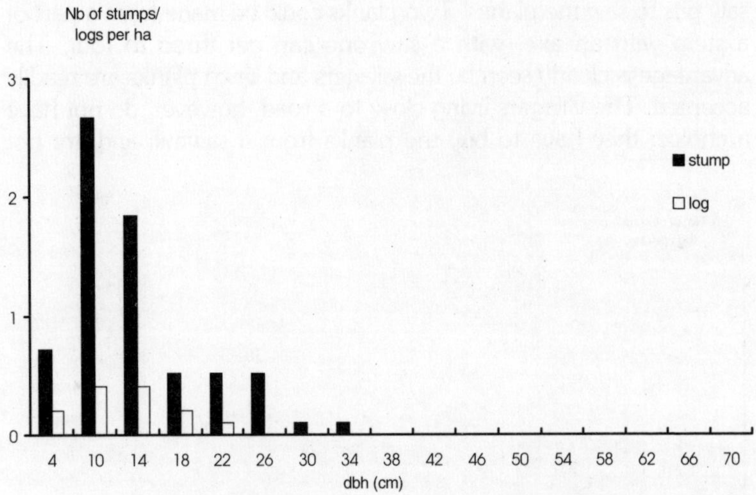

***Fig.* 5: Diameters of blue pine stumps and logs felled for poles.**

OTHERS

Blue pine is also used for furniture and a Bhutanese villager's house generally consists of several wooden boxes to store personal belongings or food, shelves for the kitchen utensils, an altar and, small tables for rituals. The size of the altar and the workmanship of the tables and altar depend on the economic wealth of the household. The very soft wood of blue pine is also used for carvings and masks. The carvings adorn temples, altars and houses. The masks are worn in ritual dances.

Non-timber forest products (NTFP)

Compared to southern regions (e.g. Wagner 1994), only a small number of NTFP is collected in Dhur. The most important NTFP is a tall bamboo species of (*Borinda grossa*), which is used to make the walls inside the houses, fences and baskets. Very important in the past, *kuan* (torchwood) is nowadays largely replaced by kerosene. Furthermore, leaves and needles are collected as bedding for domestic animals.

SOK (LITTER)

Dead leaves and needles are collected as litter by the Monpas. Women collect the litter with racks and carry it in baskets to the village. In some regions, (e.g. Chhumme) litter is collected once a year and piled up near the village. In Dhur, the litter is mostly collected when needed. Some 30 per cent of the households have *sokshing*: a patch of forest where the people have the usufructuary right to collect litter, as per the Land Act, 1978, but have no right to the trees or the land.

KUAN (TORCHWOOD)

Kuan (torchwood or candlewood) is collected from pine trees. The stem of a pine tree is damaged with a knife or an axe so that the tree is shedding resin. After some time, the wood is soaked with resin and burns readily. Kuan was widely used for lighting, as a torch, or to start the fire in the kitchen or the campfire. Nowadays, those who can afford it, replace it by kerosene which is more convenient to handle and produces less soot. The Brokpas appreciate kerosene: one or two small canisters of kerosene will provide them with lighting for the whole summer, whereas in the past, they had to carry one or two Yak loads (40-80 kg) of kuan. The soot on the stones above the places in the house where the torches were put, was formerly collected to produce ink and had to be paid as tax in kind.

OTHERS

The green needles of blue pine are used as *sang,* the locally collected incense. It is composed of fresh or dried leaves of different plants. *Sang* is offered in purification rituals in the households every morning. The Monpas mainly use fresh blue pine needles, which are readily available near the settlement. The green needles of blue pine are burned in small stoves outside the houses. The Brokpas use mostly *Rhododendron anthopogon*, *R. nivale* and *Juniperus sp.*, which they collect on their summer pastures or in the forests nearby. Green needles of blue pine are spread on the ground for some rituals or celebrations (e.g. His Majesty's birthday) as a 'green carpet'. The bark of young shoots of blue pine is peeled and the inner bark (*whad*) is chewed. The taste is sweetish and mostly children eat it. The villagers believe that regular consumption of whad in spring time prevents tuberculosis.

Impact on the structure of blue pine forests

NUMBER OF STEMS PER HA

During the winter of 1997-98, a forest inventory was carried out with a square grid of 200m × 200m. 55 plots were measured in pure blue pine forests and blue pine forests with an oak understorey. The plots were circular with a radius of 12.62 m. Trees above 8 cm dbh (dbh = diameter at breast height) were recorded, as well as all the stumps and logs, because the structure of a forest can be more accurately characterised with the help of the diameter of tree stumps and logs. Blue pine forests and blue pine forests with oak showed a similar distribution of diameters so that they could be grouped. The distribution was exponential: on a semi-logarithmic scale, the number of stems per diameter category decreases regularly with increasing diameter [*fig.* 6]. The main difference between these two forest types

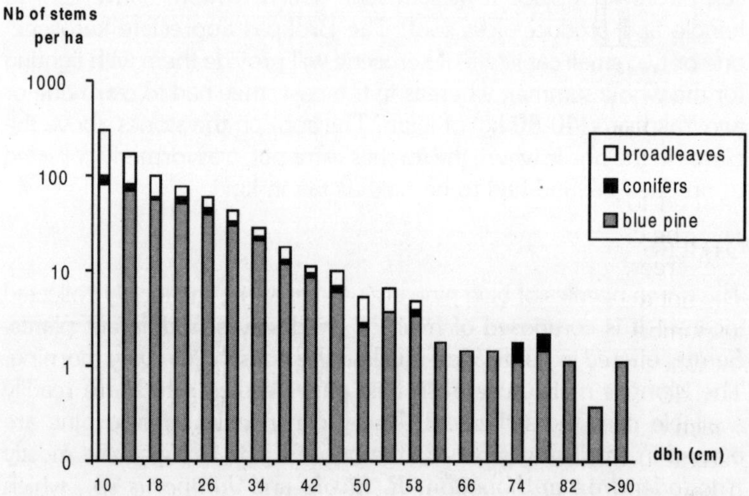

***Fig.* 6: Diameters in blue pine and blue pine with oak forests.**

were the higher number of broadleaves, mainly oak, in the latter. Oaks can grow to 30 m in height, but are generally smaller than blue pines and make up the middle layer of a forest. The forest use is not equally distributed over the whole area. Some forests were located on steep slopes, some close to settlements or footpaths, and others were located in remote areas. During the inventory, what was used in each plot was recorded. The next task was to assess the impact these forest uses would have on the diameter distribution of the trees in these plots. *Fig.*7 shows the diameters of trees in forests which

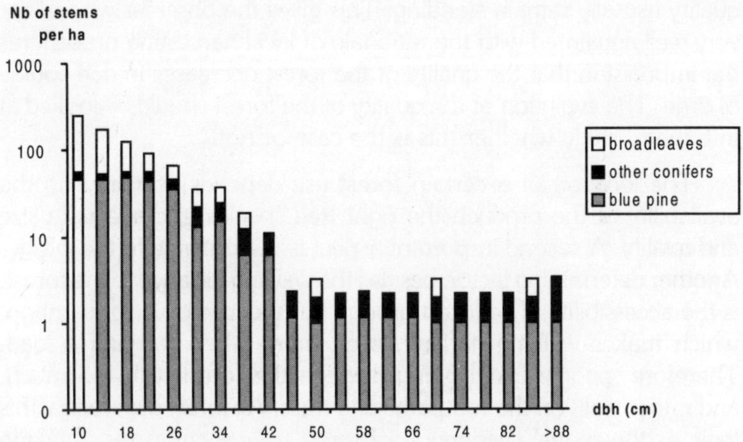

***Fig.* 7: Diameter in blue pine and blue pine with oak forests used for beams.**

are used for harvesting beams in a moderate or a more intensive way. The difference between the two graphs is obvious. In forests in which trees are harvested for beams, the distribution of the diameter does not descend in a straight line. Indeed, there is a strong drop in the number of trees between the categories of 42 cm dbh, and 46 cm dbh, and a smaller drop between those of 26 cm and 30 cm. From 46 cm dbh onwards, only one blue pine per ha is to be found in each category. The average size of the trees used for beams was presented above: it is around 30 cm dbh. In this diameter category we have the first irregularity in the diameter distribution. Those blue pines which are over 44 cm dbh and are still standing in these forests are those which were not suitable for beams. They were either forked or damaged. It is evident that the extraction of trees for beams strongly affects the structure of the forest. On the other hand, we found this use in a few forest patches only and large tracts of forests are not affected by this selected harvesting. The other forest uses did not have a significant impact on the diameter distribution, either because the uses are extensive, or similar to a thinning process in their effect.

The structure of the blue pine forest on the basis of diameter distribution had to be supplemented by further aspects such as the quality of the trees. Since the villagers choose the trees according to quality criteria (straightness, with few branches...), the trees of lower quality ususally remain standing. This gives the observer who is not very well aquainted with the rationale of local harvesting procedures the impression that the quality of the forest decreases in due course of time. The evolution of the quality of the forest should be looked at in a future study whether this is the case or not.

The location of a certain forest use depends primarily on the availability of the product: the right tree species, the adequate size and quality. A second important aspect is the distance to the village. Another determining factor, besides the walking distance to the forest, is the accessibility. The forest ground has a dense cover of bamboo, which makes walking difficult, even more when carrying a load. Therefore, people avoid going through the forest with too much, and rather walk on the footpaths and trails. All forest uses follow this logic as the spatial distribution of forest uses show in *fig*. 8 taking the example of fuelwood collection. Fuelwood collection is very intensely concentrated close to the village and around the pastures of Khamgi, where it is needed for the camp-fires of the Yak herders. It is also considerable along the ridge, since the fields and pastures are on the other side. The fuelwood is either used on the spot were it is collected or carried home in the evenings.

Rationale of use

From the above figures one can see that the villagers use distinct blue pine trees for particular products which they have in mind when they select them. For fuelwood the range of diameters is widest and unspecific. A great number of small trees are cut, but big trees are cut as well. Most probably the big trees are cut for *jarshing* (summer firewood), whereas the smaller ones are cut when the cattle herders come back home. There is a spatial distribution of the trees used for fuelwood. The bigger ones are located in the vicinity of the village. Since only a limited number of trees are allotted to each household, it is interesting for the villager to get bigger trees. Around the temporary camps of the Yak herders, smaller trees are felled since it is easier for the herders to fell and chop small to medium-sized trees

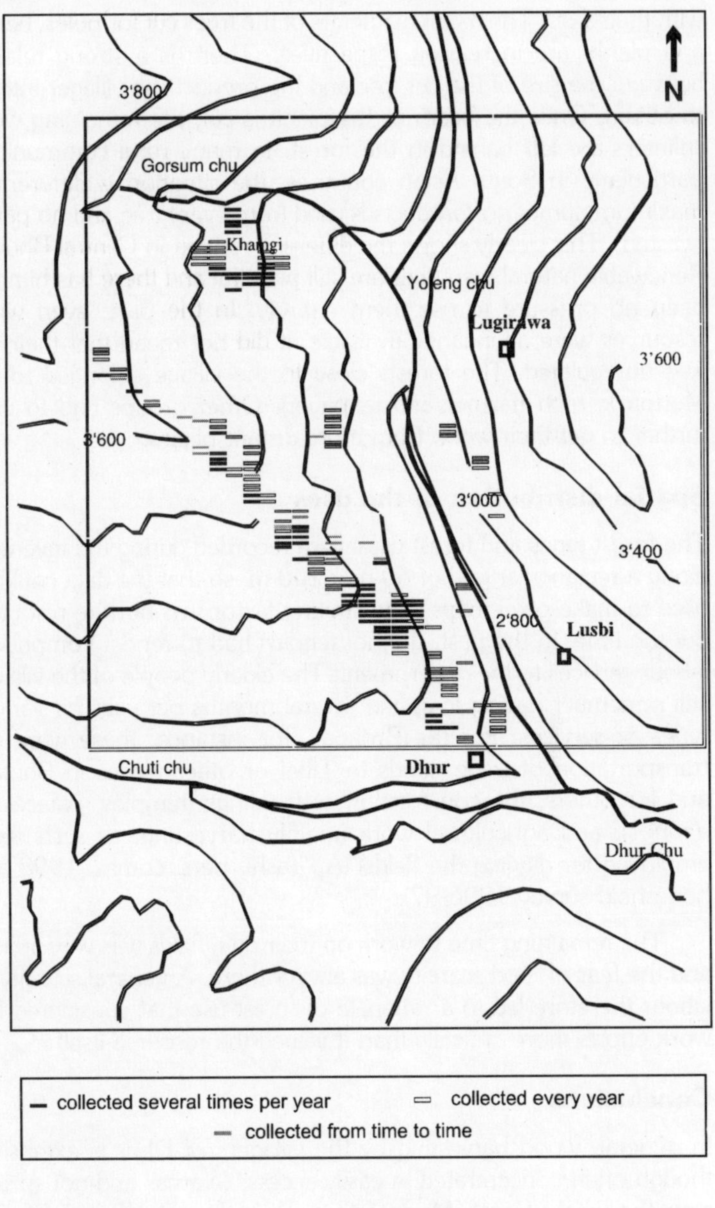

Fig. 8: **Location of fuelwood collection sites. Scale 1:50,000.**

with their axes. The mean diameters of the trees cut for poles, beams and planks are increasing respectively. There is a strong relation between the size of the cut tree and the product the villager intends to obtain. Once the beam or the plank is cut, the remaining wood splinters are left behind in the forest. In many rural communities, particularly in South Asian countries, the situation is different: a maximum number of products is used from every tree and no part is 'wasted'. This clearly shows the different context in Central Bhutan. Renewable natural resources are still plentiful and there has hitherto been no pressure to use them entirely. In the past, even when resources were abundantly available, it did not mean that their use was unregulated. The forests close to the village belonged to the Monpa *khraba* (farmer and taxpayer). Other groups had to walk further to get their wood from more distant places.

Spatial distribution of the uses

The forest types and forest uses were recorded during the inventory along a rectangular grid of 50 m × 200 m, so that the data could be used to make pixel maps. The limiting factor was not the resource, but the time. In the past, the households had to render compulsory labour services to the government. The elderly people of the villages still remember having to spend several months per year for various types of services. For the Brokpas, for instance, these were the transportation of trade goods to Tibet or other places in Bhutan; and for others, the construction of buildings (temples, palaces or *dzongs*), and agricultural work (mainly harvesting, as serfs were employed for digging the fields) (Ap'Tashi, pers. comm. 1998 and household survey 1996-97).

The remaining time to work on their own fields was very limited and the fear of food scarcity was always there. A general scarcity of labour therefore led to a rationale of forest use that considered the work efforts more critically than it valued the resource itself.

Conclusions

In general, wood harvesting by the villagers of Dhur is extensive, though often concentrated in easily-accessible areas and not spread over the whole forest. Most of these uses do not affect the forest structure with respect to tree species and diameter distribution, except in the case of blue pines used for beams. In these cases the impact on the forest structure could be clearly assessed. This research, however, did not measure the impact of the traditional forest uses on the quality of the forest. This should be examined in future studies.

The rationale of the farmers for the selection of trees differs considerably from that of the foresters. According to silvicultural and forest management theory, a stand will be thinned several times. Every time the best trees will be favoured by removing their most serious competitors. The stand will be definitively harvested after having reached its optimal annual increment, in order to get a maximum volume in the long run. The villagers, however, take only one product from each tree and choose the tree according to the product they have in mind to be made out of it. From a professional forester's point of view, they remove the best trees, before they reach their optimal volume, and spare the trees of lower commercial quality. The ultimate conclusion of this study is that the rationale of the villagers is oriented towards an optimal time management and not towards an optimum use of renewable forest resources.

References

Aris, M.,1994, *The raven crown: the origins of Buddhist monarchy in Bhutan*, London, Serindia. 160p.

Brandis, D., 1906, *Indian trees: an account of trees, shrubs, woody climbers, bamboos and palms indigenous or commonly cultivated in the British Indian Empire*, London, Archibald Constable & Co., 767 p.

Central Statistical Organization, 1994, *Statistical Yearbook of Bhutan 1993*, Thimphu, Central Statistical Organization, Ministry of Planning, Royal Government of Bhutan. 153 p.

Champion, H.G. and Seth, S.K., 1968, *A revised survey of the forest types of India,* New Delhi, Manager of Publications, 404 p.

Driem, G.van, 1993, *Language policy in Bhutan. A traditional order and the forces of change*, London, Strachan-Kiscadale.

Economic and Social Commission for Asia and the Pacific 1991: *Bhutan: Explanatory brochure*, New York, United Nations. 56p, and 2 maps.

Gansser, A., 1983, *Geology of the Bhutan Himalaya*, Basel, Birkhäuser Verlag. 181 p.

Grierson, A. J. C. and D. G. Long, 1983, *Flora of Bhutan*, Edinburgh, Royal Botanic Garden, 186 p.

Kovda, V. A. and E. V. Lobova, 1971, *Soil map of Asia*, n°4. Moscow.

LUPP, 1994, *Bumthang Dzongkhag*, Thimphu, Ministry of Agriculture.

LUPP, 1995, *Data sheet for all gewogs in Bumthang dzongkhag*, Thimphu, Ministry of Agriculture.

Polunin, O. and A. Stainton, 1997, *Flowers of the Himalaya*, Delhi, Oxford University Press, 580 p.

RGOB, 1969, *The Bhutan Forest Act*, Thimphu, Ministry of Trade, Commerce, Industries, Forest and Mines.

———, 1978, *The Land Act*, Law Committee, Thimphu.

———, 1991, *Master Plan for Forestry Development*, Department of Forestry, Thimphu.

Roder, W., O. Calvert, V. Dorji, 1992, 'Shifting cultivation systems practised in Bhutan', *Agroforestry systems*, 19: 149-58.

Roder, W., O. Calvert, V. Dorji, 1993, 'Effect of burning on selected soil parameters in a grass fallow shifting cultivation system in Bhutan', *Plant and Soil*, 149: 51-58.

Rohrbach, E., 1989, *Silvicultural research in the forestry units Dhur, Domkhar, Hurchi and in the Lame Gompa research forest (1986-1988)*, Zürich, Helvetas.

Rosset, J., 1998, *Temperate conifer forests of Bhutan. A review of forestry research activities until June, 1998*, Jakar, RNR-Research Center Jakar.

Sargent, C., O. Sargent, R. Parsell, 1985, 'The forests of Bhutan: a vital resource for the Himalayas?', *Journal of Tropical Ecology*, 1(3): 265-86.

Troup, R.S., (1986 reprint), *The sylviculture of Indian trees*, vol. 3, Dehra Dun, International Book Distributors, 1195 p.

UNDP 1996, *Bhutan: Development co-operation report 1994*, Thimphu, UNDP.

Wagner, J., 1994, *The ethno-botany of four villages in Zhemgang dzongkhag: a rapid rural appraisal*, Save The Children, Zhemgang, 31p.

Wangchuk, S., 1998, *Local perceptions and indigenous institutions as forms of social performance for sustainable forest management in Bhutan*. Forstwissenschaftliche Beiträge der Professur Forstpolitik und Forstökonomie Nr. 20, ETH Zürich, 178p.

World Bank 1994, *Bhutan. Country economic memorandum*, South Asia Region, World Bank, South Asia Country Department I.

16

Simple Participatory Forest Inventory

A key factor for sustainable forest management

Robin aus der Beek

Introduction

THE Forest Rules of His Majesty's Government of Nepal, passed in 1995, relating to the 1993 Forest Act, specify that two distinct stages need to be completed before a forest can officially be handed over to a Forest User Group by the District Forest Officer. First, the user group associated with a particular forest area needs to be formed and registered. This requires the preparation of the Constitution of the Forest User Group, mentioning the members of the group, the rules, regulations and administrative procedures relating to the use of the forest, the elected representative committee, individual responsibilities, and last but not least, the boundaries of the community forest.

The second stage is for the user group to produce an Operational Plan describing the forest management to be applied for the protection and improvement of the forest to be handed over, as well as for the fulfilment of the needs of the members of the Forest User Group. This Operational Plan also needs to be approved by both the users and the District Forest Office. The Operational Plan contains more detailed information about the present condition of the forest, and the management goals of the group.

Actually, there is considerable pressure on field staff both from the District Forest Office and from the forest users, to hand forests over as rapidly as possible. Consequently, forest areas are very often handed over to user groups with only minimal provisions being made for their effective management. Not surprisingly, in the absence of encouragement to do otherwise, user groups tend to protect rather than manage them (M. Hobley 1996).

For the last two years, the Nepal Swiss Community Forestry Project (NSCFP) has identified many potential Non-Government Organisations (NGOs), capable of supporting the Forest User Groups (FUGs) formation process in a bottom-up approach, thus achieving a complementary and synergetic effect with the activities of the District Forest Office. However, none of these NGOs had experiences in the handing over process, nor had a participatory process been defined on how to fulfil all the requirements mentioned above. These are some very crucial issues for the successful implementation of the Community Forestry Programme in the mid-hills of Nepal. With the Community Forestry Programme, the farmers have become the major actors in the planning process of forest management and its implementation, while the District Forest Offices have started to play more of an advisory and supportive role. In order to support the empowerment of the farmers, the approach strategies for the forest management practices have to be redefined, too. Traditionally, professional forest management has been carried out following scientific and technical procedures. These can hardly be followed by local farmers, who are often illiterate. Being illiterate, however, does not mean that they have no understanding of the forest. The new challenge is to define forest management practices which can be easily understood and implemented by non-professional users, without compromising on the sustainability of the productivity. The involvement of local NGOs increases the workload on the District Forest Office staff, but the involved NGOs have a great need for capacity building.

Based on these circumstances, the NSCFP Action Research Cell decided to give high priority to supporting the FUGs formation process. The main focus was to develop and test simple methods on how to define and implement, together with user groups, the socio-economically and ecologically appropriate management of forest and other tree resources. Forest management practices cannot be successful and sustainable without prior accurate assessment of the forest condition, its capacity to fulfil the demand of the users (supply of forest products) and appropriate terms of silvicultural treatment. Therefore, the method has to guarantee:

- the understanding and active participation of the concerned users.
- a short time investment.
- the use of simple instruments easily available even for FUGs in remote areas.
- an accurate estimation of every product available in the forest (fuelwood, fodder, timber, ground grass, dry and green litter, leaves, NTFPs, etc.).
- an accurate assessment of the forest condition.

Based on the assessment of the forest condition as well as on the interests of the FUG, it is possible to define the main objectives for the forest management, the silvicultural requirements, the annual harvestable products and some additional recommendations linked to the improvement of the biodiversity of the forest. For that purpose a simple and participatory forest inventory practice has been worked out, and has already been tested during the formation process of many FUGs in Dolakha and Ramechhap Districts. The purpose of this paper is to highlight some major characteristics of this simple method for participatory forest inventory for forestry technicians and forest users at the grass-roots level. All partners involved (members of the FUG, NGOs, District Forest Office staff) easily understood and applied the method with high appreciation. This has already led to a considerable improvement in the quality of the newly-formed FUGs in both districts.

The challenge

THE IMPROVEMENT OF FOREST MANAGEMENT PRACTICES

Forests are resources which, if producing the right mix of products, could contribute much to the socio-economic living standard in the villages and especially to that of the poor and women. Therefore, forest management planning should take into account the present and future value of a range of products, which could be obtained through different silvicultural options. Information on the production potentials for different products can help the FUGs to manage their forest better. Unfortunately, the reality in the mid-hills of Nepal is not moving in this direction, mainly because of a lack of knowledge linked with the appropriate way of how to improve the forest condition in order to get the highest benefit from it. Based on the experience of their elder generations, farmers mostly believe that harvesting is synonymous with forest destruction, while the total protection of the forest is considered the most efficient way to improve the forest condition.

When a forest is handed over to a user group, a management plan is usually worked out easily, focusing on a long list of measures on how to protect the forest such as to stop grazing, to control fire, to watch the forest, and restricting the silvicultural operations to the harvesting of some dead trees and branches lying on the ground. Instead of creating income-generating activities and supplying important products to the farming system, the closure of the forests handed over (or at least the reduced harvesting intensity) is seriously affecting the living condition of the farmers. The poorest ones, in particular, are often forced to sell their cattle because of a lack of fodder or private land to keep their livestock in, and they have to spend many hours a day to find fuelwood for their households. This total forest protection, however, is not an efficient way to improve the forest condition.

SUSTAINABLE SOCIO-ECONOMIC BENEFITS FROM THE FOREST

In 1997, there was no net return for the Korthali Forest User Group (a FUG in the district of Dolakha in the middle hills of Nepal), even though around 250 farmers did pruning and cleaning activities in 5 ha of their forest. The harvest which could have been made through a silvicultural first thinning, aimed to improve the future production and value of the forest. This would have resulted in an estimated gross total value of NRs. 6,000,000, and at the same time would have improved the productivity and the quality of the forest. The low harvesting intensity, the application of inadequate silvicultural operations, and limitations of knowledge on harvesting techniques, are some of the causes for the minimal benefit the FUG could achieve. This under-utilisation of plantations as well as of natural forests is common in the hill districts of Nepal (Robinson 1997).

At the same time, it can clearly be stated that fodder is a forest product of crucial importance for the farmers with respect to milk production and the maintenance of the soil fertility. However, only very few statements (mostly in terms of restricting fodder collection) can be found in the forest management plans on the potential of the forest for fodder production. Another surprising fact is that in many FUGs, where the forest could sustain a greater harvest intensity, the Operational Plan does not even consider fuelwood as a by-product of timber extraction. This happens in areas where everybody relies on it and *no fuelwood* means *no food*.

New economic potentials in forests based on the management of NTFPs and the opportunity given to local people to make sustainable use of them is raising the interest of many FUGs. However, even if the situation seems to be very appropriate to include the sustainable use of NTFPs in the forest management plans, this is not happening in most areas of the country. It seems to be clear that, even if the farmers get the right to decide about how to manage the forests handed over to them, they underestimate the production potential of the forest and thus loose a great deal of their socio-economic benefits.

EQUITY

It is not surprising that the poorest people are most affected by the imposed restrictions to use forest products after the forest has been handed over to a FUG. The richest farmers have enough private land from where they can easily get sufficient fodder, firewood and even timber for their subsistence. The situation is quite different for the poorer villagers who have little or no private land at all, and therefore have to face hard times to get those products, even if there are cases where FUGs define special conditions to improve their situation. At the same time, even if the forest has been protected for several years, and even if the FUG starts to harvest some products, the management is mainly focused on the interests of the higher castes, which is mainly for timber production and less on other products which are being much more important for the disadvantaged groups. This situation shows that the management planning did not take place in a participatory way where everybody could express their needs and interests. In order to increase the participation of disadvantaged groups in the discussion and the decision-making about the forest management practices, it is important to create a congenial and informal environment where poor people feel at ease to participate. A simple participatory forest inventory practice proved to be an excellent occasion for the poorest people to express their opinion and participate in decision-making on future forest management practices.

GENDER BALANCE

The situation described above, concerning the participation of disadvantaged groups also applies to women. Most of the work related to the forest is done by women, therefore they are probably the most knowledgeable about the opportunities and the limitations to be considered in the planning of forest management. However, even with the recent improvements in gender balance, the society is still dominated by men and, particularly in the rural areas, it is quite difficult for women to express their opinion in the public and take part in the decision-making process. During the participatory forest inventory exercise, it will be much more appropriate for women to share their knowledge and voice their expectations.

THE NEED FOR A NEW FOREST INVENTORY PRACTICE

Since the implementation of the Community Forestry Programme in Nepal the role and responsibilities of the involved actors in the definition and implementation of forest management practices have changed considerably. The traditional forest management practice has therefore to be replaced by a new one, relevant to the present situation. In Table 1 the main changes towards a participatory forest management practice are described.

The forest inventory practice described in this document does not pretend to be a new practice, but more an integration of different practices tried out in the field, or mentioned in the literature. However, it has proved to be very effective for the improvement of forest management practices as referred to in Table 1.

The simple participatory forest inventory

Two complementary steps to forest inventory are dealt with in this document. The first one focuses on the estimation of the forest productivity in terms of timber, while the second one estimates the

Table 1: Needs for a participatory forest management

Traditional forest management	*Participatory forest management*
Resource assessment and management are focused on timber production.	Resource assessment and management have to focus on the diversification of the production (timber, fodder, fuelwood, ground grass, litter, NTFPs, eco-tourism, etc.).
Inefficient forest management (forest protection rather than sustainable production).	Forest protection and utilisation must be compatible and complementary.
Complex, scientific and less flexible forest management.	The forest management must be simple, less 'scientific'. Monitoring by the users must allow them to adjust the management plan to the changes in the forest condition.
Use of scientific language, hardly understood by the farmers.	The forest management must be explained in a simple language, easy to be understood by the farmers.
Quantification is done using measures (units), which are not very familiar to the farmers (ha, metres, cubm, kg, ...)	The quantification has to be done by using measures well known to the farmers (ropani, cubft, dharni, bhari,...).
No considerations (or only very few) of conservation issues or improvement of biodiversity.	Simple, cost-neutral recommendations have to be given concerning conservation and biodiversity.
Low degree of participation of women and disadvantaged groups in the decision-making processes.	Increased participation of women and disadvantaged groups in the definition and implementation of forest management.
No collaboration between Forest Officers and farmers for sustainable forest management.	A complementary sharing of knowledge, roles and responsibilities is indispensable between farmers and Forest Officers.

available quantity of all the other products (firewood, fodder, ground grass, litter, NTFPs).

REQUIREMENTS FOR A PARTCIPATORY FOREST INVENTORY

During the FUG formation process and the preparation of the Operational Plan a wide range of information is needed. Before starting the forest inventory the following essential information has to be gathered:

- Total number of users (number of household members of the FUG).
- House-to-house visits for the need assessment of forest products (how far these can be obtained from the forest or from private land).
- Users' expectations from forest management.
- Forest boundary survey map.
- Participatory forest map with Block/Sub-Block divisions.

The collection of these data are crucial elements for forest management planning, including discussions, visits and Participatory Rural Appraisal exercises which have to be conducted prior to the forest inventory. They are essential in order to assess the socio-economic conditions of the users. Yet, it does not elaborate on the block and sub-block divisions of the forest (for more details about these topics and other details of the methods see Rai aus der Beek 1998).

The estimation of the growing stock

THE ESTIMATION OF TIMBER

For the estimation of the available amount of timber for harvesting, two different procedures can be followed:

- The relascope inventory
- The circular plot inventory

The main difference between the two procedures is the selection of the trees to be measured. In the first case the trees to be measured are selected with a simple wooden relascope (*fig.* 9), while for the second procedure the trees are measured in sample plots of 5.64 m radius (plot area = 100 sq.m = 0.2 ropani). As shown in *fig.* 9, the relascope is a very simple instrument that allows the user to have a very quick estimation of the basal area and the timber volume in the forest. The person who makes the measurement with the relascope has to stay exactly in the centre of the plot. Turning in 'clock-wise' direction, he will calculate each tree independently from its distance.

The trees to be measured are selected by holding the end of the stick of the wooden relascope below one eye and aiming through the gap 'a' at every stem at breast height. If the pole of the tree (at breast height) is wider then the opening 'a' (holding the relascope bar very close to the eye), then the tree must be measured, otherwise not (*fig.* 10).

The bigger the trees in the forest, the bigger the factor of the relascope being used can be. Relascope Factor 1 is more appropriate for trees with dbh (diameter at breast height) up to 20-30 cm, while Relascope Factor 2 is more appropriate for trees with dbh bigger than 20-30 cm.

For both inventory practices (relascope inventory and circular plot inventory), the users have to record the species name, the girth (or dbh), average height, and quality factors such as the light condition, crown form and bole quality of each selected tree. For this exercise, some instruments are required like the wooden relascope, a clinometer for the height measurement (if this is not available, even a stick of armlength can be used), a measuring tape, one calculator (if necessary this can be provided by the forest technician) and one wooden peg or stone to permanently mark the center of the plot.

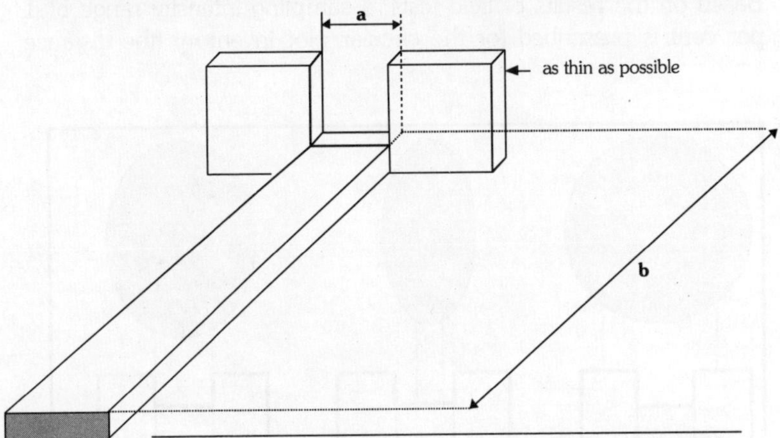

	Relascope Factor 1	Relascope Factor 2
a	1.0 cm	1.4 cm
b	50 cm	50 cm

***Fig.* 9: Description of a simple wooden relascope.**

During the measurement, the forest technician explains and discusses with the farmers some silvicultural practices, which will be later described in the forest management plan (selection of trees to be cut with reference to the light condition and the form of the crown, as well as on the bole quality, pruning, thinning, singling, etc.). Local leaders will learn the practice very fast and often want to be the main actors of the measurement activity. In order to be sure of the motivation of these people, but also to involve the women and the disadvantaged people in the measurement, the leaders are asked to teach the other participants how to use the instruments. The role of the technician as facilitator is a crucial factor to do the measurement in a really participatory way.

The data analysis (not further described in this presentation) will give an estimation of the available timber per hectare, from which 1-3 per cent can be harvested every year. Once the estimations have been done hectare-wise, they can be extrapolated according to the total area of the forest, and must finally be converted into local measures (instead of cubic metre/ha, indicate cubic feet/ropani).

SAMPLING INTENSITY

Based on the results of field tests, a sampling intensity range of 1 per cent is prescribed for the circular plot inventory (the distance

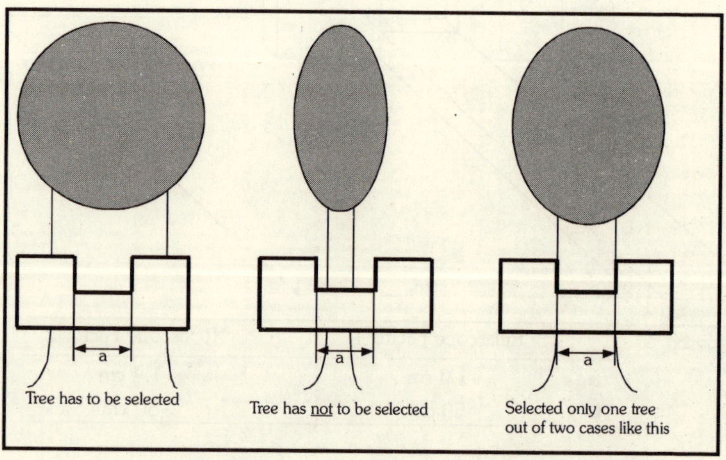

***Fig.* 10: Selection of the trees with the wooden relascope.**

between the plot centres is 100 m), while for the relascope inventory the sampling intensity is defined as shown in Table 2, depending on the forest area and the Relascope Factor used.

Table 2: Number of sampling plots and distances between the centres of the plots for the relascope inventory

	Relascope Factor 1		*Relascope Factor 2*	
Forest area	*Plot /Ha (No./Ha)*	*Distance (m)*	*Plot / Ha (No. / Ha)*	*Distance (m)*
< 4 ha	1.6	79	2.4	65
4-8 ha	1.5	82	2.3	66
8-16 ha	1.4	84	2.1	69
16-32 ha	1.2	91	1.9	73
32-64 ha	1.0	100	1.6	79
> 64 ha	0.8	112	1.2	91

EVALUATION OF THE REGENERATION CONDITIONS

In order to ensure the sustainability of the forest management, it is important to look at the conditions for regeneration, especially in old forests which need to be regenerated quite soon. The forest technician has to first discuss with the farmers the importance of the natural regeneration, and then to explain the meaning of the measurement practice. To estimate the regeneration, all seedlings and saplings between 0.25 m and 2.00 m height are counted by species in a circular plot of 1.78 m radius (circle area = 10 sq m = 0.02 ropani). Similarly, for the assessment of the established regeneration, all trees taller than 2 m and smaller than 8 cm in diameter found within a circular plot of 2.82 m radius (circle area = 25 sq m = 0.05 ropani), are recorded, taking the name of the species, girth (or the diameter) and height.

The centre of the plot must be the same for both the circular plot for the timber estimation, as well as for the centre for the Relascope inventory. Once the plot centres have been defined, the users, with a rope the length of the respective plot radius, turn around the plot centre in a clock-wise direction and count or measure all trees that fulfil the mentioned criteria.

THE ESTIMATION OF FIREWOOD

For the estimation of the available quantity of firewood, a circle of 5.64 m radius (circle area = 100 sq.m = 0.2 ropani) is demarcated by placing the users on the border (circumference) of the circle itself (*fig.* 11). The centre of the circular plot must be the same centre used for the estimation of timber. The farmers (women and men from different castes) are then asked to visually estimate the total amount of fuelwood available inside the circle, assuming that all trees and shrubs would be cut (total amount of fuelwood). The wood of taller trees, having good boles that could be used as timber or for other purposes, should not be included in the estimate. After estimating the total amount of fuelwood, the farmers assess the amount of fuelwood they would collect inside the circle, taking sustainability into consideration, and indicate how much time should pass before they could harvest in the same plot again. Once the estimations have been done for the circle area, they can be easily extrapolated for the total forest area and mentioned in the Operational Plan by using the local measures (bhari/ropani, or dharni/ropani)[1].

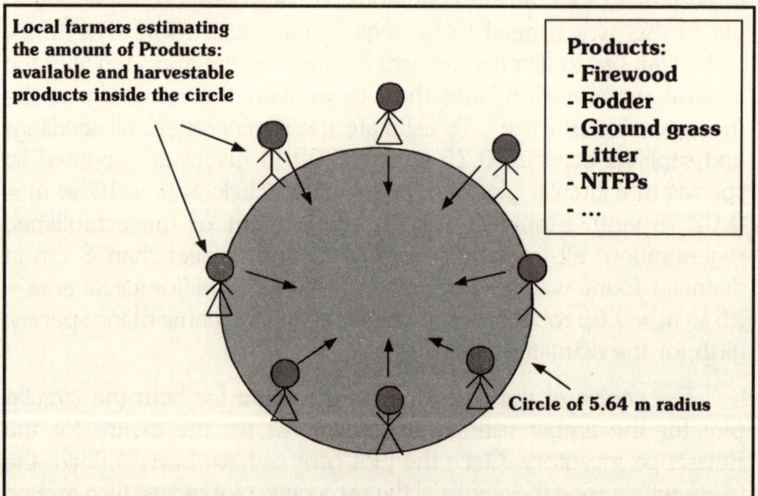

***Fig.* 11: Circle plot for the estimation of the harvestable products**

1. 1 dharni = 2.5 kg and 1 bhari = 40 kg.

ESTIMATION OF THE HARVESTABLE AMOUNT OF FODDER, GROUND GRASS AND LITTER (GREEN LITTER OR DRY LITTER)

The same circle formed to estimate the available fuelwood is used to assess the amount of other products like fodder, ground grass and litter ready for harvesting, and the time required between two harvesting operations. Once the farmers have made these assessments by themselves, it will be no problem to include them in the Operational Plan. It will also include the interests of women and disadvantaged groups in the process of forest management. The management will no longer be focused on timber production (according to the interests of men, particularly from higher castes) and lower caste people are not likely to suffer from unjustified harvest limitations. Similar to the estimation of fuelwood, the measured units of these products have to be in dharni/ropani, or bhari/ropani).

LIST OF THE AVAILABLE NON-TIMBER FOREST PRODUCTS (NTFP)

It has been shown that the potential for income-generating activities by harvesting NTFPs in Dolakha and Ramechhap Districts can be very interesting, especially because it can create job opportunities for women and disadvantaged groups (Lafranchi 1997, Paudel & Rosset 1998). Yet, very few FUGs have given attention to the management of NTFPs so far. Therefore, it is important to include the list of available NTFPs, and even of those NTFPs which are not available at present but could be introduced and managed in a sustainable way, in the Operational Plan. During the inventory, the farmers must be asked to list all NTFPs they have seen in the forest. If they are thinking of commercialising these products, they have to quantify the available amount and to define a sustainable harvest regime.

SITE QUALITY

The site quality is an important factor for the growth of trees and other plants. Normally no special equipment is available for local farmers to measure the soil composition. However, some basic information is required, such as slope, aspect, soil depth, organic matter content, drainage, etc. The soil character helps to define the site quality and site index which will help to project the growth rates of the bio-mass.

BIODIVERSITY

Depending on the needs and demands of the community, the functions of the forest and its importance can vary considerably and even change with time. A successful forest management plan must be based on a detailed analysis of all the functions that the forest is expected to accomplish, and has to find the most appropriate way to reconcile functions which might even be contradictory, like production and protection. In this context, the role of the forest in terms of the conservation of biodiversity is no exception. Practices like the collection of dead wood and leaf litter from the forest floor which destroy the micro-habitat of invertebrates, mosses, fungi and lichens, and their continued removal, may lead to reduced biodiversity and degradation of the forest site. Some Operational Plans mention a need to remove 'unwanted species'. These species may not be common or economically important, but their loss still affects biodiversity. Even dead trees can play an important role for the biodiversity of the forest, providing adequate nesting sites for many bird species, as well as for providing a habitat to many insects which are important for the diet of birds and other animals of the forest. Small creeks can be very important because of the rich flora and fauna of those ecosystems, and should therefore be protected. This kind of observation and subsequent recommendation in the Operational Plan can be a small but important contribution to the biodiversity of the forest.

Conclusions and recommendations

- On average, the full measurement of one sample plot as described in this document takes one hour. This might be considered a considerable time investment, but if sustainable forest management is the aim, only such a forest inventory can provide the required basic information.
- The present simple participatory forest inventory gives equal importance to every forest product. By including every forest product in the forest management practice, a great contribution is made to the improvement of the socio-economic conditions in remote rural areas.
- The forest inventory gives equal participation opportunities to women and disadvantaged groups in order to share their

knowledge and to express their needs. Once their interests have been considered during the forest inventory, they will be more ready to become an integral part of the forest management system.

- The forest management practice can only be participatory, if the language of the farmers and the local measurement units are used.
- The harvesting intensity of only 1 to 3 per cent of the available timber is relatively low compared to what the forest can produce. However, it is already much higher than previously mentioned in the Operational Plans and accepted by the District Forest Offices. The monitoring of the forest condition will reveal how to adjust this initial estimation according to the forest development.
- The forest inventory draws a clear picture of the forest condition and its need for improvement. Therefore, it has to be considered as an essential step to the development of a sustainable management plan.
- Even if the forest can supply many products to the farm households, it might not be able to cover all the needs of the users. Through the forest inventory, the users can easily understand the limitations of their forest, and therefore will be aware of the importance of increasing the productivity on their private land through agroforestry.

References

Bayerisches Staatsministerium, 1979, Hilfstafeln für die Forsteinrichtung, Bayerisches Staatsministerium für Ernaehrung, Landwirtschaft und Forsten, Bayern, Germany, 383 pp.

Hobley, M., 1996, Participatory Forestry: The Process of Change in India and Nepal. Rural Development Forestry, Study Guide 3, Overseas Development Institute, London, England, 337 pp.

Lafranchi, S., 1997, The economic potential for Forest User Groups by sustainable production, transformation and commercialization of some Non-Timber Forest Products, in the hill Districts of Dolakha and Ramechhap, Nepal, Nepal Swiss Community Forestry Project, Action research Cell, Technical Note No.2, Kathmandu, Nepal, 110 pp.

Paudel, D. and Rosset, Chr., 1998, What Hanuman brought was not only 'Jadibuti...'. Nepal Swiss Community Forestry Project, Action research Cell, Technical Note No. 1, Kathmandu, Nepal, 99 pp.

Rai, C. and aus der Beek, R., 1998, Simple Participatory Forest Inventory and Data Processing; Guidelines for the Operational Plan Preparation of Forest User Groups, Nepal Swiss Community Forestry Project, Action Research Cell, Technical Note No.3, Kathmandu, Nepal.

Robinson, P., 1997, Report to SDC: Consultancy Mission to Nepal Swiss Community Forestry Project (17.02.1997 to 07.03.1997), t.311 Nepal 50, t.751 - 416 - BRP, Bern, Switzerland, 16 pp.

1. Peasants on agricultural terraces in the nearby of Piple VDC (Ramechhap, Nov. 1999)

2. Agroforestry in Budikhoriya VDC (Ramechhap, Dec. 1999)

3. Part of the village of Bhudha Bhimsen VDC (Dolakha, Nov. 1999)

4. View on a pat of Bhudha Bhimsen from its community forest (Dolakha, Dec. 1999)

5. Group interview with members of the executive committee of Budhikhoriya FUG under a Pipal (*Ficus religiosa*) tree (Ramechchap, Dec. 1999)

6. Hamlet surrounds by terraces near Taamaa Koshi river (Ramechhap, Nov. 1999)

7. Non timber forest products among other goods sold on a market (Ramechhap, Dec. 1999)

8. PRA exercise (participatory mapping) in Bhudha Bhimsen (Dolakha, Dec. 1999)

9. Bhimsen Temple in the community forest of Bhudha Bhimsen (Dolakha, Dec. 1999)

10. Ritual performed by a Brahman in the nearby of a tea-shop in Bhudha Bhimsen (Dolakha, Nov. 1999)

11. Individual interview and use of PRA tools; (Dolakha, Nov. 1999)

12. Dongaria Kondh village

13. Dongaria Kondh after slashing the undergrowth in a forest

14. Swidden fields after burning

15. Sacred grove in Mayurbhanj

16. Dongaria man

17. Dongaria woman

18. Castor plants in swidden fields

19. Market at Belaghar in Central Orissa

20. Kuttia Kondh village

21. Kuttia Kondh man making a plough

22. Kuttia Kondh woman with her child

23. Kuttia Kondh priest

24. Village gardens in Kendujhar District

25. Village in Kendujhar District

26. Village in Central Orissa

27. Village, fields and forest in Central Orissa

28. Degraded land in the semi-arid area of Central Rajasthan

29. Well preserved semi-arid area in Central Rajasthan

30. Desert in Western Rajasthan

31. *Brokpa* (Yak herder) kitchen

32. Bhutanese ritual dancer performing in a religious festival

33. Bhutanese child making butter

34. Bhutanese family with monk

35. Women weaving in Kurji village, Central Bhutan

36. Typical house of well-to-do farmers in Central Bhutan

37. Chhoekor valley in Central Bhutan

38. Trongsa in Central Bhutan

39. High altitude lake with Yak (*Bos grunniens*) in Northern Bhutan

40. Chomolhari Mountain (7315 m) in Northwestern Bhutan

Glossary

Ajan	:	Muslim call to prayer (adhan)
Brahmana	:	highest caste (priests) in the Hindu social hierarchy
Crore	:	10 million
Fetishism	:	worshipping of objects believed to have magical power
Gautam baudhitva	:	becoming a Buddha
Gram Panchayat	:	village committee
Gram Sabha	:	village council
Kharif	:	summer crop
Khatedar	:	tenant of government land
Khesra forest	:	not demarcated protected forests
Krishi Vigyan Kendra	:	farmer education centre
Kshatriya	:	second highest caste (warriors) in the Hindu ritual hierarchy
Mahila Samiti	:	women's club
Naxalites	:	militant political activists
Non-scheduled areas	:	regions where no special regulations with regard to SC/ST apply
Padmabhusan	:	order of the Indian Union
Padmashree	:	order of the Indian Union
Paharia	:	mountain dweller
Palli Sabha	:	village assembly

Panchajanya	:	conch-shell
Panchayati Raj	:	village government system
Pradhan Panch	:	elected chairman of the village council
Puja	:	celebration of a religious ritual
Rabi	:	winter crop
Rishi	:	saint
Sarpanch	:	village headman
Scheduled areas	:	regions where special regulations with regard to SC/ST apply
Shivaratri	:	Shiva's night — Hindu festival
Sundhi	:	learned man
Swidden	:	plot of land occasionally slashed and burned for cultivation
Tehsil	:	revenue district
Totemism	:	social structure ordered by clan symbols
Zaheera	:	sacred grove

Index

Adhikari,J., 284

Agriculture, for food procurement, 69-71, 77-78

Agroforestry,
- associations and interdependence, 200-01
- biomass production-oriented strategy, 203-04
- concept and importance, 197-98
- definition, 195
- development and management of, 195-208
- in dryland agriculture, 201-02
- incentives, 205-08
- indigenous, 203-04
- land-use system, 196
- pastoralist groups influence, 199
- perennials and annuals complementary usage of, 204-05
- policy formulation, 205-08
- social factors influence, 198
- socio-economic advantages, 201-02
- systems, 202-03

Agropastoral groups, influence on agroforestry, 199

Arabari pilot project, 147

Aravalli Afforestation Project, 182

Aravallis, forest management in, 175-76

Aris, M., 295

Arrowroot, 116

Baitarani river, 29-30, 32, 56

Ballabh, Vishwa, 197

Basnet, Sitaram, 254

Baumer, Michel, 197

Beek, Robin aus der, 311

Bharara, Laj Pal, 195, 197

Bhattachan, K.B., 265

Bhise, S.N., 173

Bhitar-Kanika Wildlife Sanctuary, 55

Bhuinya tribe, 76

Bhutan,
- blue pine,
  - forests, 294
  - impact on structure of, 304-06
  - rationale of use, 306-08
  - spatial distribution of uses, 308
  - traditional use, 298-302
- economy, 296
- forest types, 293-94
- geology and climate, 289
- infrastructure, 297
- institutions, 297
- land use, 292-93

NTFP in, 302-03
population, 294-96
social and economic conditions, 294-97

Bhutan Forest Act 1969, 297-98

Bhuyan, Dharanidhar, 2

Bhuyans of Kendujhar, 2

Biodiversity, in Juangpirh, 55-66

Birhor tribe, 113-14

Bista, 284

Biswal, Ragunath, 155

Blue pine,
construction wood, 299-302
beams, 298-300
planks, 300-01
poles, 301-02
fuelwood, 289-99
traditional use of, 298-99

Bose, N.K., 59, 85

Brahmani river, 29

Brandis, D., 294

Brokensha, D., 36

Bu (irrigated land), 280

Budhikhamari FPC, 161-64, 172

CFM committees, 87-89, 99-101

CHIPKO movement, 147

Calvert, O., 293

Campbell, J.Y., 82

Central Arid Zone Research Institute (CAZRI), 197

Central Bhutan, forest structure in, 289-309

Central Statistical Organization, 295, 297

Chambers, Robert, 36, 82

Champion, H.G., 57, 294

Char seeds, 111

Charan (pasture land), 281

Chhatia FPC, 154-56, 172

Choudhury, B.P., 55

Churia Forest Development Project, Nepal,
aerial photograph interpretation and forest inventory design, 224-25
area, 213
biophysical features, 213
community and private forestry, 216, 218-19
demographic, socio-economic and livestock features, 213-14
experiences with,
community forestry, 219-22
private forestry, 222-23
forest policy and institutional set up, 230-33
forest situation, 214-16
forest strategy, 212-13
forest technical aspects, 233-34
fuelwood supply demand scenario, 217
philosophy of, 211-12
problem statement, 216, 218-19
recommendation, 228-34
socio-economic features, 225-28
firewood demand, 225-26
livestock, 226-28
timber demand, 225-26
subwatershed management approach, 223

Committee on Forest and Tribals in India report, 83

Common property resources (CPRs),

177, 205
Communal hunting, 74
Community Forest Management (CFM), 132
Community Forest Management Programme, 285
Community Forestry (CF), 236-37, 239, 242, 246, 248, 252-55, 257, 260-62
  in Terai, 211-34
Community Forestry Programme, Nepal, 312, 316
Cottage industries, 167
Crop farmer groups, 199

Dalton, E.T., 68
Das, R.P., 13
Dasapalla, rebellion in, 2
Dash, J., 69
Deforestation, consequences of, 35
Directive Principles of State Policy, 2
District Rural Development Agency (DRDA), 190
Driem, George van, 296
Dryland farming, 116-18
Duerr, W.A., 122

Elwin, V., 14, 60, 76, 105
Ethnobotany,
  indigenous methods of classification, 43-46
  research methods of, 41-42
    origin of clans, 42
    origin of natural objects, 41-42

Federation of Community Forest User Groups in Nepal (FECOFUN), 233, 254
Fernandes, W., 13, 19
Fichtenau, Jürgen, 211
Fisher, R.J., 235, 238, 282
Food,
  concept of, 67
  procurement modes, 68-72
    agriculture, 69-71, 77-78
    hunting, 68-69
    socio-cultural lifestyle and, 71-72
    traditional mode, 68

Forest,
  access to, 14
  agroforestry and, 195-208
  community forestry, 211-34
  development project, 211-34
  estimation of growing stock, 318-24
  geographical distribution patterns, 228-30
  Indian culture aspect of, 13-14
  indigenous knowledge, 15-18, 126, 130
  inventory practice, 316-25
  joint forest management, 131-41
    people's institutions and, 173-92
    studies on, 145-72
  minor forest produce, *see*, Minor Forest Produce (MFP)
  NTFP, *see*, Non-timber forest products (NTPF)
  participatory management of natural resources of, 235-63, 311-25
  resource distribution, 229-30
  sustainable management, 311-25

sustainable socio-economic benefits from, 314-15
traditional forest uses in Central Bhutan, 289-309
tribes within regional and national political setup, 18-21, 24
user groups, 235-63

Forest Act of 1927, 146

Forest Act of 1993, Nepal, 311

Forest Act of 1999, Nepal, 255

Forest Department, 4, 18, 74, 89, 121, 130, 133, 135, 137-38, 140-41, 148-49, 151, 155-61, 163-64, 166-67, 175, 177, 180-81, 184, 186-92
institutional change in, 189
Forest Development Committee, 231
Forest Development Corporation, 179
Forest dwellers,
forest management and, 122
forest resource use by, 124
Kendujhar district, 28
MFP and, 106-13
regional and national political setup, 18-21, 24
socio-cultural life of, 28

Forest management, 21-23, 122-31
among Tamang of Southern Kavre, 265-88
community participation in, 148-49
definition of, 122-23
forest policies, 176-83
forest settlement and working plans, 177-78
forest-dwelling communities and, 122-23, 140-41
historical background of, 174-75
in Aravallis, 175-76
indigenous system, 126
institutional arrangements for, 175-76
institutional framework, 130-31, 135
joint forest management, 131-41
knowledge basis, 129-30
manipulation of forest, 125-29
NCA and, 178-79
natural resource management systems and, 125
needs, 123-24
participatory approach in, 179-83, 316-25
practices, 313-14
sustainable management, 311-25

Forest Nationalization Act, 1957, Nepal, 284

Forest Policy 1952, 176-77

Forest Policy 1988, 179-83

Forest produce collection, 73

Forest Protection Committees (FPCs), 142-46, 181-82
networking of, 191-92

Forest user groups (FUGs), 311-15, 317-18, 323
access to, 252-53
decision making in, 259-61
disadvantaged groups, 240
dynamics of, 247-48
ethnic and caste composition of, 245, 247-48
factors that influence participation in, 257
fund utilisation, 255-56, 261
informal discussions on, 258
institutional actors, 246
institutions, 239-40

management of natural resources, 240
participants, 248-55
access to FUG, 252-53
community forest significance, 253
on FUG fund and its utilisation, 255-56, 261
perception of forest, 251-52
protection of CF against misuse, 255
satisfied needs and consented investments, 253-54
socio-cultural profile, 248-51
participation, 255-56, 262
participatory management of natural resources with, 235-63
perception of forest, 251-52, 257
resource management system, 246-47
selection of case studies, 241
socio-cultural status, 244
study in mid-hills of Nepal, 235-63
data analysis, 243-44
information collection techniques, 242-43
methodology, 240-44
objective and framework, 237
problem and background, 235-36
recommendations of, 262-63
results and analysis, 244-61
Forest villages, 4

Gadgil, M., 13
Gansser, Augusto, 292
Gautam, 282-83
Giesch, Christina, 289
Gilmour, D.A., 235, 238, 282
Grierson, A.J.C., 294
Guha, R., 13

Hermeneutic contexts, 15
Hill Bhuinya tribe, 29
Hill Kharia tribe, 115-16
Hivale, S., 105
Hobley, M., 241, 312
Honey production, 115-16
Hota, Raghunath, 151
Human resource development, agroforestry development and management through, 195-208
Hunter, W.W., 68
Hunting, for food procurement, 68-69
Hunting and gathering tribes, 113-16, 127-28

ICAR, 196
ICRAF, 196
Indian Forest Act of 1865, 147
Indian Forest and Conservation Acts, 133
Indian Forest Law 1865, 20
Indigenous forest management, 126, 130
Indigenous knowledge (IK),
research in, 23
scope of implementation of, 15-18
system and practice, 36-40
Indigenous knowledge system and

practices (IKSP), 36-40

Indira Jungle Suraksha Mahila Sangha, 165

Ingles, A., 242

Integrated Subwatershed Management Plan (ISWSMP), 223

Integrated Tribal Development Agency (ITDA), 18, 29

International Code of Botanical Nomenclature, 57

Irrigation farming, 116-18

Jackson, B., 242

Jagir system in Rajasthan, 174-75

Jamunalia FPC, 158-61, 172

Jena, Mihir Kumar, 81-82, 85-86, 115-16

Jodha, N.S., 197, 206

Johnson, S., 132

Joint forest management (JFM), 4-5, 20, 22, 131-41, 87, 101
- absence of legal authority and legal recognition, 186
- case studies, 150-72, 184-92
  - Budhikhamari FPC, 161-64, 172
  - Chhatia FPC, 154-56, 172
  - Jamunalia FPC, 158-61, 172
  - Katalposi FPC, 156-58, 172
  - Mahapada FPC, 150-54, 172
  - problems and suggestions, 166-67
  - Purikhunta FPC, 164-66, 172
- change needed in marketing and enhancing income, 186-87
- community participation in, 148-49
- conflict resolution mechanisms, 188
- confusion over boundaries, 186
- developmental needs and, 187
- follow-up of shortcomings of resolution, 188-89
- forest official involvement, 167
- impact and possibilities of, 192
- in Rajasthan, 183-84
- institutional change, 189-92
- legal framework of, 185-87
- participatory approach in, 179-83
- plantation on wastelands, 166
- planting vs. regeneration, 185
- policy, 168-71
- policy implementation, 149-50
- resolution, 5, 131
- security, 167
- silvicultural practice, 187-88
- studies on, 145-72
- to degraded and barren forests, 183-84
- women's participation, 167

Joseph, S.J., 78

Juang Development Agency, 78

Juang of Kendujhar, 2

Juang tribe,
- concept of distinguishing plant species, 45-46
- cultivated food plants of, 51-53
- culture, 32-33
- ethnobotany, 40-42
- food procurement by, 68-72, 74-75
- forest produce collection, 73-74
- IKSP, 36-40
- livelihood, 33-34, 46
- MFP by, 35-36
- marriage, 72

natural resources use, 38-39
of Juangpirh, 32-53
plant conservation knowledge, 39-40
plants in life of, 60-66
shifting cultivation by, 34-36, 46
socio-cultural lifestyle, 71-72, 76-79
species differentiation by, 43-46
wild food plants of, 50-51

Juangpirh, Orissa,
biodiversity in, 55-66
climate, 56-57
forest produce in, 85
geography, 56
Juang of, 32-47
man-nature relationship in, 27-53
material and method, 57
medicinal plants of, 47-50
moist deciduous forest, 58-60
NTFP in, 81-102
relevant plants, 60-66
semi-evergreen forests, 57-58
vegetational types, 57-60

Kalidas, 8

Kapoor, R.P., 182

Kapp, W.W., 77

Katalposi FPC, 156-58, 172

Kendujhar district, Orissa,
climate and rainfall, 31, 56-57
forests, 30
geography of, 28-29, 56
grazing lands, 30-31
minerals, 30
population, 29
road communication, 31
structure and relief, 29
surface drainage, 29-30

Kharka (high altitude pasture land), 281

Kissling-Näf, I., 239

Kondh, Anu, 2

Kondh tribe, 111, 116-19

Kovda, V.A., 292

Lac, 111

Lafranchi, S., 241, 323

Lai, Chun K., 196

Land Ceiling Act, 206

Land Reform Act 1964, Nepal, 284

Lévi-Strauss, C., 42

Lobova, E.V., 292

Long, D.G., 294

MASS, 87

Mahant, Gorachand, 162

Mahant, Parvati, 164

Mahapada FPC, 150-54, 172

Mahapatra, L.K., 117-18

Mahua flowers and seeds, 110

Malhotra, S.P., 197

Man and Forest project, 8-9, 20

Man-nature relations, of Juang of Juangpirh, 27-53

Mankidia tribe, 113-16

McDougal, C.W., 73-75, 77

Medicinal plants, 8

Meghaduta, 8

Mehta, A.S., 173

Ministry of Environment and Forests (MoEF), 180

Minor Forest Produce (MFP), 35, 81, 83-84
- *char* seeds, 111
- collection of, 105-08
- Lac, 111
- *Mahua* flowers and seeds, 110
- multiple use of habitats, 112-13
- *Myrobolans*, 109
- *Sal,* 108-09
- *Siali* leaves, 111-12
- Tamarind, 110-11
- *Tassar*, 111
- tribal lifestyle and, 106-13

Mishra, K.C., 162

Misra, S., 59

Mixed farming groups, 199

Mohandungri village, JFM in, 184-92

Mohanty, K.K., 60

Mohanty, R.P., 77

Mohapatra, P.K., 60

Mooney, H.F., 57

Mrang (marginal land), 281

Munda, Birsa, 2

Murdock, G.P., 79

Myrobolans, 109

National Commission on Agriculture (NCA) 1976, 178-79

National Forest Policy 1988, 4, 82, 131

Natural resource management system, and forest management, 125

Nayak, Laxman, 2

Nepal,
- community forestry in, 211-34
- environmental perception and forest management among Tamang of, 265-88
- forest development project in, 211-34
- forest user groups socio-cultural study in, 253-63
- participatory forest inventory in, 311-25

Nepal Swiss Community Forestry Project (NSCFP), 238, 240-41, 249, 312

New Brahmanas, 3

Non-governmental organizations (NGOs), 16-17, 21-23

Non-timber forest products (NTFP), 16, 19, 180-81, 186-87, 302-03, 315, 317, 323
- collection of, 105
- economic and biological potential of, 86-99, 102
  - in Badampahar area, 90-92
  - in Bisoi area, 93-94
  - in Jashipur area, 95-97
- in Similipal hills of Orissa, 81-102
- livelihood and, 99-101
- policy, 82-84
- sources, 99-101, 135
- subject-policy on procurement and trade of, 83

ORMAS, 6

Orissa,
- biodiversity in, 55-66
- forest cultures, 15
- forest management policy, 5, 21-23, 146-47, 168-71
- forest use among tribes of, 105-19
- JFM in, 131-41, 145-75

Juang tribe, *see,* Juang tribe
NTFP in, 81-102
man-nature relationship in, 27-47
plants found in, 62-66
tribal forest life in, 13-24
tribal population in, 2

Orissa Forest Act 1972, 169

Orissa Village Forest Rules 1985, 5

Ostrom, E., 254

Ottiger, Nadja, 121

Padhi, Sailabala, 78

Palit, Subhabrata,124, 130, 133

Participatory forest inventory,
estimation of growing stock, 318-24
biodiversity, 324
firewood, 322
harvestable amount of fodder, ground grass and litter, 323
NTFP, 323
regeneration conditions, 321
sampling intensity, 320-21
site quality, 323
timber, 318-20
needs for, 317
requirements for, 317

Participatory Forest Management (PFM), 132, 179-83, 316-25

Pastoral groups, influence on agroforestry, 199

Pastoralists, 128

Pathak, A., 21

Pathi, P., 113-14, 145

Patnaik, Girija Bhusan, 1

Patnaik, Kamala Kumari, 105, 115

Patnaik, Nityananda, 27, 59-60, 115

Paudel, D., 323

Paudel, H.L., 240

People's institutions, 190

Plant conservation,
by restriction, 39-40
plants associated with totems, 39
through worship and magico-religious belief, 40
tribal knowledge of, 39-40

Poffenberger, M., 132, 150, 153

Polunin, O., 294

Private forestry, and community forestry, 216, 218-19, 222-23

Programme of Community Forestry (PCF), 236

Protection of Wild Life Act, 74

Puhan, N.C., 155

Purikhunta FPC, 164-66, 172

Quit India Movement (1942), 2

Radcliffe-Brown, A.R., 77

Rajasthan,
geography, 173-74
JFM programme in, 183-92
participatory approach in forestry in, 180-83

Rajasthan Forest Act, 1953, 176

Rajasthan Land Reforms and Resumption of Jagir (Amendment) Act 1954, 175

Rath, Alok, 67

Ratha, S.N., 79

Ravindranath, N.H., 124, 126, 135

Regional Forest Development Plan, 231-32
Revenue Department, 175
Rhi (forest), 281
Rio declaration, 212
Rio summit, 23-24
Robinson, P., 314
Roder, W., 293
Rohrbach, Ernst, 294
Rosset, Christian, 235, 323
Rosset, J., 292
Rout, S.P., 68
Rural Litigation and Entitlement Kendra, 132

SAGUN, 266
Sal, 108-09
Samuells, E.A., 60
Santal revolts, 2
Saoras tribe, 111, 116-19
Sarain, Madhu, 137
Sargent, C., 292
Sargent, O., 292
Satkosia Gorge, 55
Saxena, N.C., 197
Schmithüsen, F., 238-39
Seeland, Klaus, 13-14, 36, 235
Senapati, Sahoo, 85
Seva Mandir, 183-84, 187, 189, 191-92
Shah, Mahendra Bikram, 284
Shah, Prithvi Narayan, 283
Shankarnarayan, K.A., 197
Sharma, B.D., 13
Sheth, S.K., 57
Shifting cultivation, 3, 116-18, 128
Shifting cultivators, 128
Shiva. S.C., 13
Sho Pakha (roofing grassland), 281
Siali leaves, 111-12
Silvicultural practices, need for change, 187-88
Silvicultural technique, 121, 134-36
Similipal Biosphere Reserve, 55, 84, 86, 89
Similipal National Park, 84-85
Singh, B., 74
Singhbhum, Santal revolts in, 2
Social forestry, 178-79
Social Forestry Programme, 178-79
Social Science and Development Research Institute (SSADRI), Bhubaneswar, 57
Stainton, A., 294
Sustainable forest management, simple participatory forest inventory for, 311-25
Swaminathan Foundation, 78
Swidden cultivation, 69-70, 73-74, 78

Talai Forest Protection Committee, 185
Tamang, Buddhi, 265
Tamang, P.B., 265
Tamang of southern Kavre, Nepal,
birds story, 273-74
classification of,

land topography, 278-80
land, 280-81
mechanism, 276-78
plants, 282
wildlife, 282
community forestry, 285-87
cosmology, 267-70
empirical knowledge, 276-78
environment perception and forest management among, 265-88
ethnogeography song, 272
love song, 272
myth of origin, 267
plant's story, 273
power politics and forest management, 282-85
resources cultural meaning, 274-76
animals, 276
earth, 275
moon, 275
plants and trees, 276
sun, 275
wind, 276
rituals, 270-72
songs, 272
stories, 273

Tamarind, 110-11

Tassar, 111

Tejwani, K.G., 196

Tewari, D.D., 82

Tribals,
Constitution of India and, 2
development, 4
dryland farming, 116-18
ecology as way of life, 113-18
forest use among, 105-19
forest/habitats denudation and, 3
hunters and gatherers, 113-16
irrigation farming, 116-18
life in Orissa, 13-24
lifestyles, 14-15, 106-13
role in Freedom Movement, 2
shifting cultivation, 3, 116-18
vocational education to, 4-5
women education, 4

Troup, R.S.,294

UNDP, 196

UNEP, 197

USAID, 196

Upland cultivation, 74

Valley-bottom cultivation, 74

Vana Samrakshana Samiti (VSS), Orissa, 135, 145, 151-53, 155-60, 163, 171

Village Forest Protection Committees, 135-36, 145

Wangchuk, S., 289, 294

Warren, D., 36

Wastelands, plantation on, 166

Werner, O., 36

Whara (rain-fed land), 281

Wiersum, K.F., 122-23, 126

Wild Life (Protection) Act, 1972, 169

Yonjan, Amrit, 265